PROGRESS IN BIOMASS CONVERSION

VOLUME 5

Academic Press Rapid Manuscript Reproduction

PROGRESS IN BIOMASS CONVERSION
VOLUME 5

Edited by

DAVID A. TILLMAN
Envirosphere Company
Bellevue, Washington

EDWIN C. JAHN
College of Environmental Sciences and Forestry
State University of New York
Syracuse, New York

1984

ACADEMIC PRESS
(Harcourt Brace Jovanovich, Publishers)
ORLANDO SAN DIEGO NEW YORK LONDON
TORONTO MONTREAL SYDNEY TOKYO

COPYRIGHT © 1984, BY ACADEMIC PRESS, INC.
ALL RIGHTS RESERVED.
NO PART OF THIS PUBLICATION MAY BE REPRODUCED OR
TRANSMITTED IN ANY FORM OR BY ANY MEANS, ELECTRONIC
OR MECHANICAL, INCLUDING PHOTOCOPY, RECORDING, OR ANY
INFORMATION STORAGE AND RETRIEVAL SYSTEM, WITHOUT
PERMISSION IN WRITING FROM THE PUBLISHER.

ACADEMIC PRESS, INC.
Orlando, Florida 32887

United Kingdom Edition published by
ACADEMIC PRESS, INC. (LONDON) LTD.
24/28 Oval Road, London NW1 7DX

ISSN 0192-6551

ISBN 0-12-535905-5

PRINTED IN THE UNITED STATES OF AMERICA

84 85 86 87   9 8 7 6 5 4 3 2 1

CONTENTS

Contributors	vii
Preface	ix
Contents of Previous Volumes	xi
Hydroprocessing of Biomass Tars for Liquid Engine Fuels *Ed J. Soltes and Shih-Chien K. Lin*	1
Fuel Characteristics of Wood and Nonwood Biomass Fuels *Amadeo Rossi*	69
Factors Influencing Dilute Sulfuric Acid Prehydrolysis of Southern Red Oak Wood *J. F. Harris, R. W. Scott, E. L. Springer, and T. H. Wegner*	101
The Energy Costs of Increased Organics Recovery for Chemical By-Products in Kraft Pulp Mills *W. J. Frederick, Jr.*	143
Microeconomic Approaches to Biomass Fuel Pricing *E. C. Lesnick, Jr.*	163
Fuel Characteristics of Selected Species of Beached Logs in Southeastern Alaska *W. Ramsay Smith and Richard O. Woodfin, Jr.*	203
An Assessment of the Costs and Benefits of Recovering Logging Residue for Energy Use *Ellen J. Hall*	217

Review of Biomass Gasification Technology 263
   Kenneth L. Tuttle

Subject Index 281

CONTRIBUTORS

Numbers in parentheses indicate the pages on which the authors' contributions begin.

W. J. *Frederick, Jr.* (143), Departments of Chemical Engineering and Forest Products, Oregon State University, Corvallis, Oregon 97331

Ellen J. *Hall* (217), Envirosphere Company, Bellevue, Washington 98004

J. F. *Harris* (101), U.S. Department of Agriculture, Forest Service, Forest Products Laboratory, Madison, Wisconsin 53705

E. C. *Lesnick, Jr.* (163), EBASCO Business Consulting Company, Mountain View, California 94041

Shih-Chien K. *Lin* (1), Department of Forest Science, Texas A & M University, College Station, Texas 77843

Amadeo *Rossi* (69), Envirosphere Company, Bellevue, Washington 98004

R. W. *Scott* (101), U.S. Department of Agriculture, Forest Service, Forest Products Laboratory, Madison, Wisconsin 53705

W. Ramsay *Smith* (203), College of Forest Resources, University of Washington, Seattle, Washington 98195

Ed J. *Soltes* (1), Department of Forest Science, Texas A & M University, College Station, Texas 77843

E. L. *Springer* (101), U.S. Department of Agriculture, Forest Service, Forest Products Laboratory, Madison, Wisconsin 53705

Kenneth L. *Tuttle* (263), U.S. Naval Academy, Annapolis, Maryland 21402

T. H. *Wegner* (101), U.S. Department of Agriculture, Forest Service, Forest Products Laboratory, Madison, Wisconsin 53705

Richard O. *Woodfin, Jr.* (203), Pacific Northwest Forest and Range Experiment Station, USDA Forest Service, Portland, Oregon 97208

PREFACE

Price stability and even price declines have characterized the fossil fuel markets over the past few years. This stability has resulted in decreasing general interest in such areas as synthetic fuels and biomass energy. The consequence of this decrease in interest has been reduced research activity in the area of alternatives to oil. The price stability in fossil fuels, particularly oil, was caused in large part by the recent recession. That same recession also resulted in decreasing investment in biomass-fired systems. The only major countervailing force combatting the decline in biomass energy interest and activity has been the Public Utilities Regulatory Policies Act (PURPA), whose cogeneration provisions have stimulated some activity.

The stability of fossil fuel prices may be more apparent than real, however. Events in the Iran–Iraq war, with bombings of Kharg Island and threats to block the Straits of Hormuz, illustrate how fragile price stability and supply tranquility may be. Recognizing that this energy supply situation is indeed fragile, many researchers and engineers continue to pursue biomass problems and opportunities. Their activities have extended beyond the premium wood fuels to include a host of agricultural materials. Today, corncobs, cotton gin trash, and a wide variety of other materials may be considered fuel. While some researchers have developed these supplies, others have focused on the economic issues of how to price materials that once were wastes and that now have some positive economic value. Still others are examining ways to make more fuel more available and in more convenient forms.

Volume 5 of *Progress in Biomass Conversion* reflects some of the current research in the biomass area. The chapters of this volume are oriented toward a presentation of a cross section of current activity. Consequently, the volume includes such disciplines as chemistry, mechanical engineering, and economics. In presenting such papers, Volume 5 seeks to represent the multidisciplinary nature of the biomass community.

If eternal vigilance is the price of peace, then eternal inquiry into alternatives is the price of energy stability. Such inquiry must include investigations in the laboratory, in engineering and design, and in systems operation. Without diligent research, we may again leave ourselves vulnerable to the control of resources by countries whose interests are dissimilar to our own.

CONTENTS OF PREVIOUS VOLUMES

VOLUME 1

Living Resources and Renewing Processes: Some Thoughts and Considerations
    Ingemar Falkehag
Wood Fuel Use in the Forest Products Industry
    R. L. Jamison
The Economic Values of Wood Residues as Fuel
    David A. Tillman
Pyrolysis of Wood Residues with a Vertical Bed Reactor
    J. A. Knight
Methanol from Wood: A Critical Assessment
    R. M. Rowell and A. E. Hokanson
A Survey of United States and European Practices for Recovering Energy from Municipal Waste
    James G. Abert and Harvey Alter
The Silvicultural Energy Farm in Perspective
    Jean-Francois Henry

Subject Index

VOLUME 2

Logging Residue as an Energy Source
    John B. Grantham and Jack O. Howard
Genetic Improvement of Forest Trees for Biomass Production
    Bruce Zobel

Wood Fuels Consumption Methodology and 1978 Results
    G. F. Schreuder and D. A. Tillman
Sugar Stalk Crops for Fuels and Chemicals
    E. S. Lipinsky and S. Kresovich
Acid-Catalyzed Delignification of Lignocellulosics in Organic Solvents
    K. V. Sarkanen
Environmental Considerations in Wood Fuel Utilization
    William D. Kitto
Wood Fuel Preparation
    W. Ramsey Smith

Subject Index

VOLUME 3

Energy from the Forest
    Edwin C. Jahn
Chemistry of Pyrolysis and Combustion of Wood
    Fred Shafizadeh
A Comparison of European and U.S. Combustion Systems Using Biomass
    Kalevi Leppa
Alternative Cogeneration Systems Employing Biomass as Fuel: An Incremental Analysis of Heat Rates
    David A. Tillman, Ronald F. Schnorr, and John W. Sale
Consumption of Wood Fuels in the United States 1971–1980
    Charles H. Norwood and Walter L. Warnick
Evaluation of New Concepts in Biomass Fiber Field Processing and Transportation
    Peter Schiess and Keith Yonaka
A Comparison of Biomass and Coal as Feedstocks for Synthetic Fuels
    Larry L. Anderson
Biomass Fuels for Energy Security: A Policy Statement Concerning National Needs and Opportunities
    David G. Palmer

Subject Index

VOLUME 4

Lipid Crops for Chemicals and Fuels
    E. S. Lipinsky and S. Kresovich
Lignin Utilization: Potential and Challenge
    Stephen Y. Lin

Adhesives from Natural Resources
   Ed J. Soltes and Shih-Chien K. Lin
Formation of NO and Particulates during Suspension-Phase Wood Combustion
   Robb M. Winter, James R. Clough, and David W. Pershing
Wood Energy Use in the Wood Products Industry: What the Data Show
   Alberto Goetzl and Susan Tatum
Advances in Chemical Pulping Processes
   William O. Aho
Making the Best Energy Use of Wood
   David A. Tillman
A Predictive Model for Stratified Downdraft Gasification of Biomass
   T. B. Reed and M. Markson
Small Scale Industrial Biomass Systems
   Norman Smith and Richard C. Hill
Biomass Augmented Ocean Thermal Energy Conversion Systems
   Malcolm S. Jones, Jr.

Subject Index

HYDROPROCESSING OF BIOMASS TARS
FOR LIQUID ENGINE FUELS

Ed J. Soltes
Shih-Chien K. Lin

Department of Forest Science
Texas Agricultural Experiment Station
Texas A&M University System
College Station, Texas

I.   INTRODUCTION 2

II. BIOMASS RESIDUES AND TARS 3

    A. Biomass Residues 3

    B. Generation of Tars 4

    C. Tar Compositions 4

    D. Tar Corrosivity 9

III. TAR HYDROPROCESSING 21

    A. Catalysts 28

    B. Hydrogen-Donor Solvents 29

    C. Hydroprocessing Conditions 30

IV. COMPOSITION AND UTILITY OF HYDROPROCESSED TARS ... 30

    A. Pine Waste Pyrolysis and Corn Cob Gasification Tars 30

    B. Other Hydroprocessed Tars 30

    C. Fuel Comparisons 50

    D. Engine Evaluations of Experimental Fuels 50

V. PROCESS CONSIDERATIONS 60

    A. Idealized Process Flowsheet 60

    B. Other Process Considerations 62

    C. Process Definition 64

VI. CONCLUSIONS 64

I. INTRODUCTION

No biomass conversion process will be implemented on a commercial scale until a slate of useful products is identified and developed. Since biomaterials are structurally and chemically complex, some product selectivity is a prerequisite to implementation, but will come at some expense, either to process engineering and/or to product engineering. Biomass thermochemical conversion produces tars, chars and gaseous products. Depending on reaction parameters, some product selectivity can be obtained in thermochemical conversion: gasification offers high yields of a relatively simple gas mixture (Soltes, 1980a, 1983a). Pyrolysis, however, yields all three products, often in similar yields. The tars have found little utility because of chemical complexity and poor chemical and physical properties (Soltes, Wiley, and Lin, 1981). For this reason, the bulk of biomass pyrolysis research today is in process engineering, in the definition of pyrolysis parameters which result in good yields of less complex tar products (Antal, 1983; Chatterjee, 1981; Diebold and Scahill, 1983). In research of this nature, however, products will be those defined by feedstock and process parameters (e.g., high yields of levoglucosan via vacuum pyrolysis). Under such conditions, products may not be immediately marketable (again, e.g., levoglucosan), and will still have to be chemically changed to permit contributions to the utilization of biomass an an alternate source of commodity chemical and energy products.

As an alternative to this approach, one can first identify large volume products which need chemically complex raw materials for their synthesis. The suggestion here is that postpyrolysis product engineering, as opposed to pyrolysis process engineering, may offer a route to commercial implementation. For example, motor fuels are mixtures of many chemical entities, and motor fuel properties cannot be fully served by less complex mixtures. Tars, as complex chemical mixtures, may be suitable feedstocks for conversion into motor fuels. For such products, one can accept the chemically complex tar product from simpler "slow" pyrolysis systems as a fuel and chemical feedstock, instead of entertaining single products via complex pyrolysis processing operations.

Previous research in this laboratory had resulted in the identification of catalysts suitable for the catalytic hydroprocessing of pine waste pyrolysis tar to a hydrocarbon fuel feedstock (Soltes, 1982, 1983b). Some insights into the mechanisms by which hydrocarbons are produced from pine pyrolysis tar suggested that the processes under development had application to a wider selection of tars and tarry condensates, not only of biomass pyrolysis, but also of biomass gasification. Reported here is research directed towards establishing to what extent the hydroprocessing of biomass thermochemical conversion tars may be a universal route to the production of functionally similar fuel and chemical products from physically dissimilar biomass feedstocks (Soltes, 1983c).

II. BIOMASS RESIDUES AND TARS

A. *Biomass Materials*

Eight types of biomass residues were obtained from local sources. These were pine wood chips, pecan shells, sugarcane bagasse, peanut shells, corncobs, rice hulls, cottonseed hulls and pine plywood trim. With the exception of pine chips, these materials are residues in the sense that little use other than combustion is the general experience. The wood chips were obtained in larger quantities, primarily to check and set gasification/pyrolysis parameters in the reactor used to generate tars. All materials were allowed to equilibrate to air-dry moisture levels before use. Although more biomass materials could have been entertained, the list was limited to those shown because it was judged adequate to demonstrate the versatility of the processing developed, and because the smaller list allowed a more comprehensive study of materials used.

B. Generation of Tars

Tars used in this research were produced in a modified gasification/pyrolysis reactor from the eight biomass residue materials. In addition, two tars produced elsewhere were subjected to processing operations: the Tech-Air tar produced from pine sawdust and bark in a 100 ton per day demonstration plant in Cordele, Georgia, and a tar produced in gasification of corncobs by DeKalb Ag Research in DeKalb, Illinois.

In gasification, where tar is undesirable, tar is still produced in as much as 10% yield (for example in updraft gasifiers), but generally in lower than 3% yield (Soltes, 1983a). For pyrolysis, tar yields are variable, but generally about 25% of dry feed, such as reported for the pine waste tar derived in the Tech-Air pyrolysis process (Knight, Bowen and Purdy, 1976). Tars were produced here in a reactor originally a downdraft gasifier, but modified for updraft gasification/pyrolysis. Tar production was approximately 10 to 20 weight percent. Some 1 to 10 kg of tar was collected from each biomass feed using an expansion condenser, a design adapted from one reported for third-world application (Tatom et al., 1980).

C. Tar Compositions

The chemical and physical characteristics of the biomass tars produced in the gasification/pyrolysis reactor are given in Table I, along with those for the Tech-Air tar.

Earlier analysis of the Tech-Air tar (Elder, 1979; Elder and Soltes, 1979a; 1979b; 1980) indicated that solvent extractable phenolics comprise approximately 13% of the tar in which phenol, cresol, quaiacol, dimethyl phenols, alkyl quaiacols and eugenol are the major components. Research towards the use of these phenolics for adhesives is described elsewhere (Soltes and Lin, 1983b). Nonextractable, high molecular weight phenolics account for another 20% of the tar. Tar also contains volatile organic acids (see Table II), as well as a variety of neutral components (Soltes and Elder, 1981). Very few components are present in greater than 1% concentration (see Table III). Figure 1 is a chromatogram for the GC volatiles of the Tech-Air tar, most components of which are the phenolics.

TABLE I. Chemical and Physical Characteristics of Biomass Tars

	Tech-Air	Wood chips	Pecan shell	Sugar-cane bagasse	Peanut shell	Corn cob	Rice hull	Cotton hull	Plywood trim
Elemental analysis[a]									
C, wt. %	65.79	54.34	68.37	55.94	60.89	58.34	64.62	67.91	74.08
H,	7.13	6.31	5.70	7.00	7.21	6.32	6.66	7.47	9.10
O,	27.08	39.18	25.32	36.75	28.87	34.30	25.28	23.02	16.49
N,	–	0.14	0.61	0.31	3.03	1.04	3.44	1.60	0.33
Heat content, MJ/kg[a]	23.18	24.32	23.93	22.39	27.65	24.88	24.26	23.86	24.61
Specific gravity[a]	1.14	1.22	1.11	1.15	1.10	1.26	N/A	N/A	1.10
Ash content[a]	1.03	0.22	0.20	0.57	0.77	1.32	0.68	3.16	0.12
Water content[b]	10.0	6.4	13.0	19.6	15.0	3.1	5.0	5.0	10.6

[a] Not corrected for water content.
[b] Water content obtained by azeotropic distillation of tar with toluene at 80°C.

TABLE II. Relative Abundance and Percentage Composition of Volatile Acids in Tech-Air Pine Waste Pyrolysis Tar

	Formic acid	Acetic acid	Propionic acid	Peak No.3[a]	Butyric acid	Peak No.5[a]	Isovaleric acid
Relative abundance	17.9	100.0	13.47	0.50	3.66	1.00	1.00
% Composition	0.32	1.70	0.24	0.01	0.06	0.02	0.02

[a] Peak Nos. 3 and 5 were not identified in gas chromatography, but exhibit acidic properites similar to the other organic acids.

TABLE III. Components in Tech-Air Pine Waste Pyrolysis Tar in Greater Than 1% Concentration

Compound	% of Tar	Compound	% of Tar
4-methyl guaiacol	4.0	acetic acid	1.7
guaiacol	2.7	dimethyl phenols	1.6
anhydroglucose	2.1	4-ethyl guaiacol	1.5
m,p-cresols	2.1	5-methyl furfural	1.2
phenol	1.7	o-cresol	1.0

All gas chromatograms given in this chapter (with the exception of the GC-mass spec runs) were derived using a Tracor 560 gas chromatograph with split injection) into a 30-meter DB-5 bonded phase fused silica capillary column. Temperature programming for the tars was held 5 minutes at 30°C then to 280°C at 3°C per minute, for the hydroprocessed tars was held 5 minutes at 30°C then to 280°C at 4°C per minute. Signals detected by a flame ionization detector were stored in, and reconstructed with the Chromatography Applications Package of an IBM 9000 Computer System.

Compositions were determined by additional gas chromatography-mass spectrometry analyses using the same type of fused silica capillary column but in a Hewlett-Packard 5985A instrument. These runs were subcontracted to Radian Corporation in Austin, Texas. Temperature programming for these runs was as given in the legend for the chromatograms. Identification of mass spectra was by comparison of spectra with those in the literature. Computer-assisted component identification was not used.

Figure 2 is the GC-mass spec chromatogram for the Tech-Air pyrolysis tar. The mass spectra generated in this run were identified and used to label the peaks in Figure 1. Additional GC-mass spectrometry on tars or their fractions were used to label the chromatograms of other tars which follow (Figures 3 through 10). Note the similarities in the volatiles composition of these tars. This is not totally unexpected since thermal degradation processes cannot distinguish between the cellulose, hemicellulose and lignin composition of, for example wood vs. corncobs. Thus, although the physical forms of various residues are different, tars produced via thermochemical processes exhibit many similarities in chemical composition.

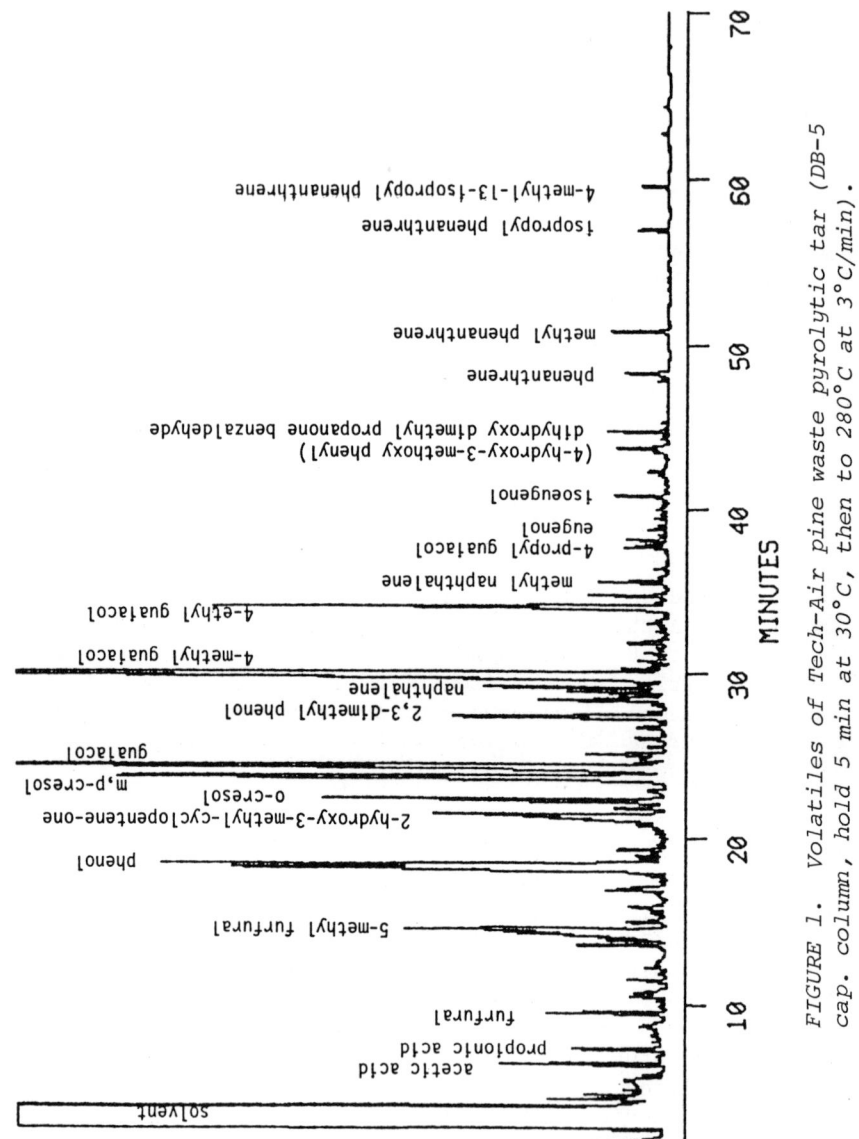

FIGURE 1. Volatiles of Tech-Air pine waste pyrolytic tar (DB-5 cap. column, hold 5 min at 30°C, then to 280°C at 3°C/min).

Gas chromatography is suitable for the volatiles of tars and their hydroprocessed products, but cannot be used for their nonvolatile components. Gel permeation chromatography can be used for the separation of functionalities in tar, both volatile and nonvolatile. GPC can be followed by analysis of the separated fractions by GC and HPLC (Sheu et al., 1984). GPC separations were performed on a Waters Associates Model ALC/GPC 202 liquid chromatograph equipped with a Model R401 refractometer. Four 10nm micro-Styragel columns (7.8mm I.D., 30 cm. long) in series were used. Tetrahydrofuran, refluxed and distilled over sodium wire in a nitrogen atmosphere, was used as solvent.

GPC of the Tech-Air tar results in the trace given in Figure 11. Fraction 1 at lower retention volumes is the polymeric content; fraction 2, larger molecules (C14 to C44 size if hydrocarbons); fraction 3, phenolics; and, fraction 4, aromatics. The composition of the aromatic fraction was surprisingly very complex, a trace of the GC-mass spec run for this fraction is given in Figure 12. Components identified by GC-mass spec are given in Table IV.

D. *Tar Corrosivity*

Tars produced via biomass pyrolysis or gasification are corrosive. Although this fact has long been known, it is interesting that there is no work published on the nature of this corrosivity, or the identification of the acids responsible. ASTM G31-72 was employed for this work, wherein mass loss and microscopic examination are reported as principal measures of corrosion (Lin, 1978).

Six different types of metals (aluminum, naval brass, mild steel, 304 stainless steel, 316 stainless steel and copper) were disk-shaped to a diameter of 1.50 inches and a thickness of 0.125 inches. Each disk had a hole with a diameter of 0.312 inches located near the edge. These specimens were precleaned by degreasing with bleach-free scouring powder, rinsing with a suitable solvent, and then finishing with abrasive paper. The dried specimens were then weighed to an accuracy of 0.5 mg (ASTM G31-72).

The apparatus to measure corrosion rates of the various metals consisted of a 1000 ml resin flask, a reflux condenser, a supporting device, a nitrogen tank, and a heating device. Test solutions included the Tech-Air pyrolysis tar, its steam distillate, and an acid solution which contained the same percentage of formic and acetic acid as found in the pyrolysis tar (see Table II). The metal disks were suspended in these three solutions for a period of 7 days at 70°C.

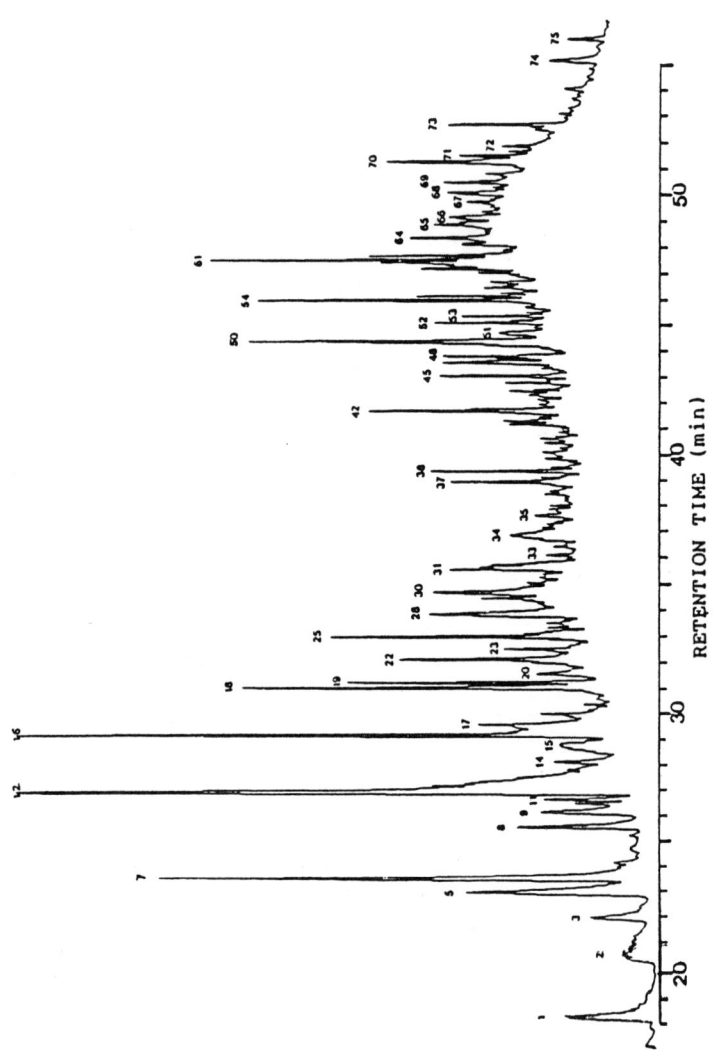

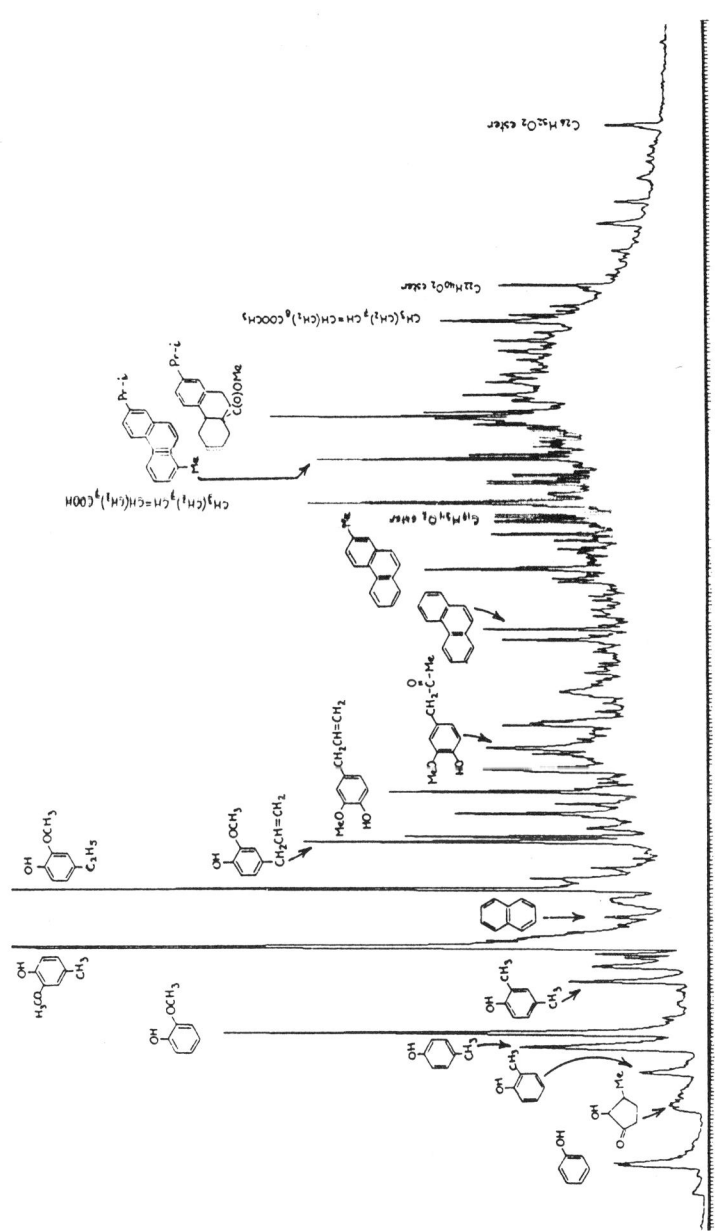

FIGURE 2. GC-Mass Spec Chromatogram of Tech-Air pyrolytic tar. Conditions: 30 meter DB-5 fused-silica capillary column, on-column injection, hydrogen carrier gas at 1 ml per min., temperature programming 30°C to 100°C at 3°C per minute then to 280°C at 6°C per min.

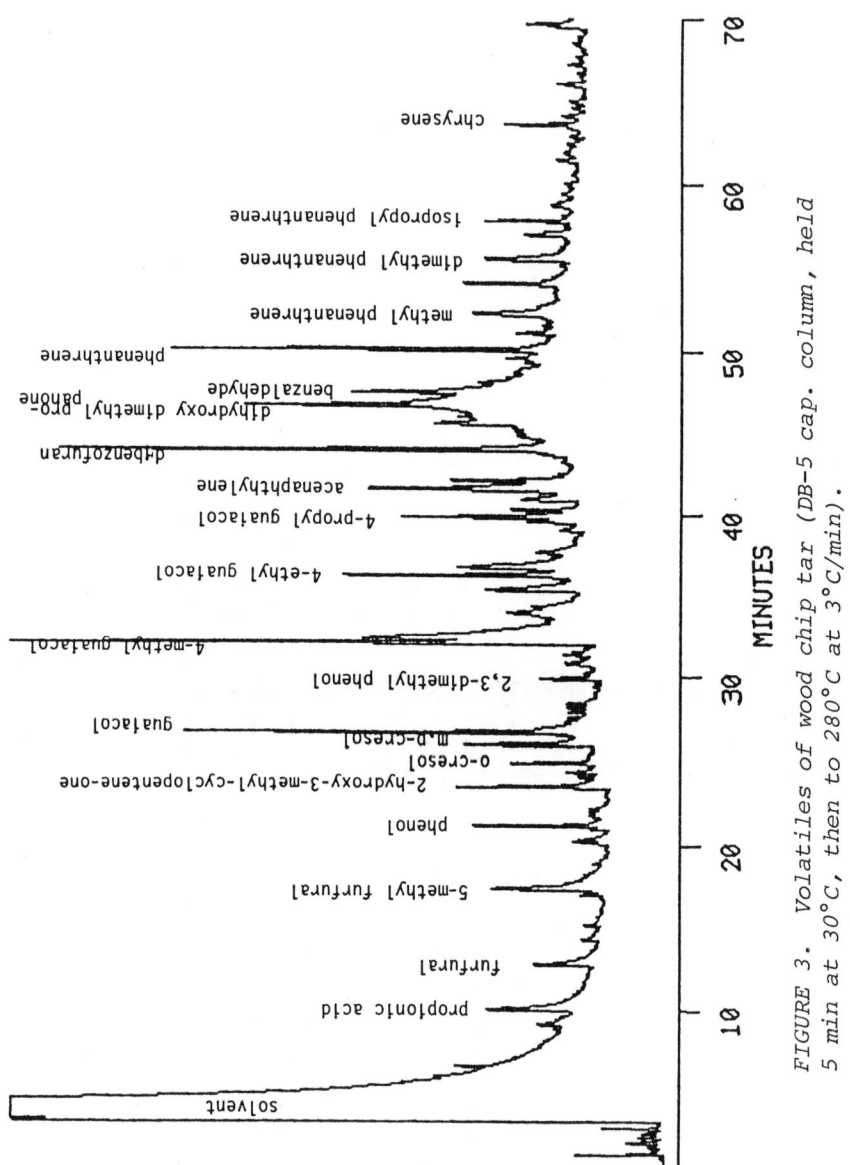

FIGURE 3. Volatiles of wood chip tar (DB-5 cap. column, held 5 min at 30°C, then to 280°C at 3°C/min).

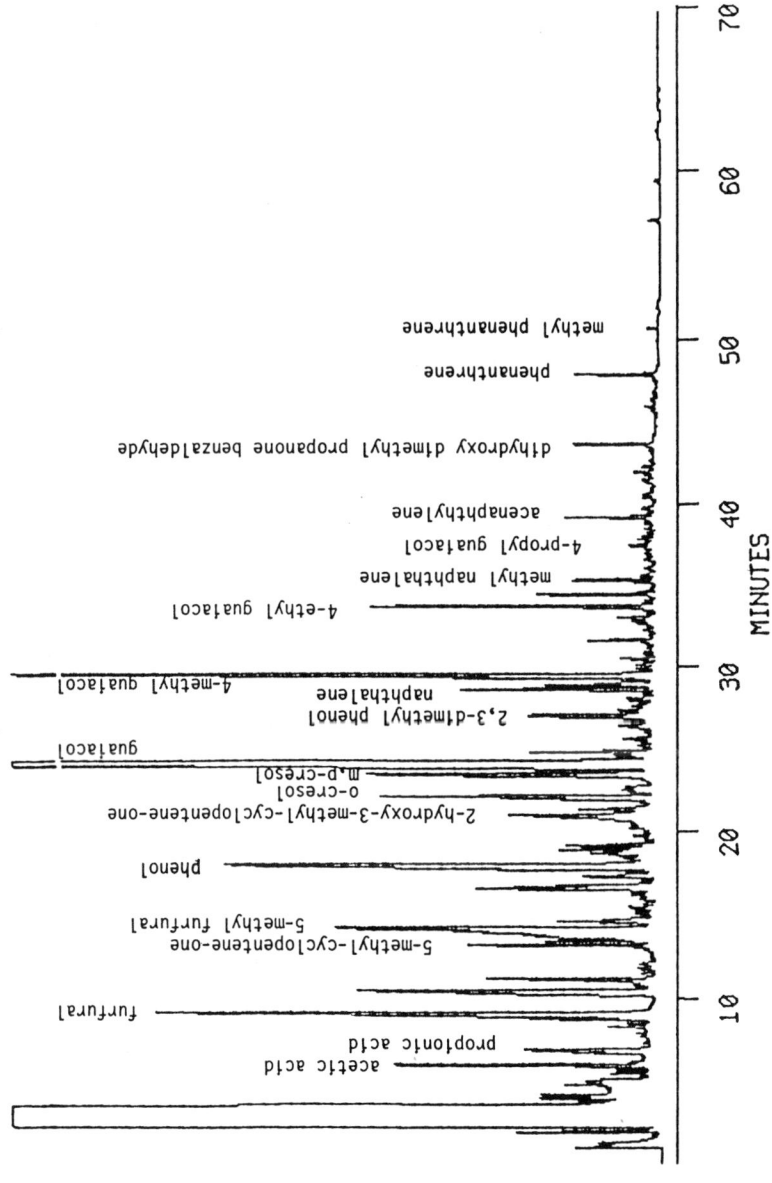

FIGURE 4. Volatiles of pecan shell tar (DB-5 cap. column, held at 30°C, then to 230°C at 3°C/min).

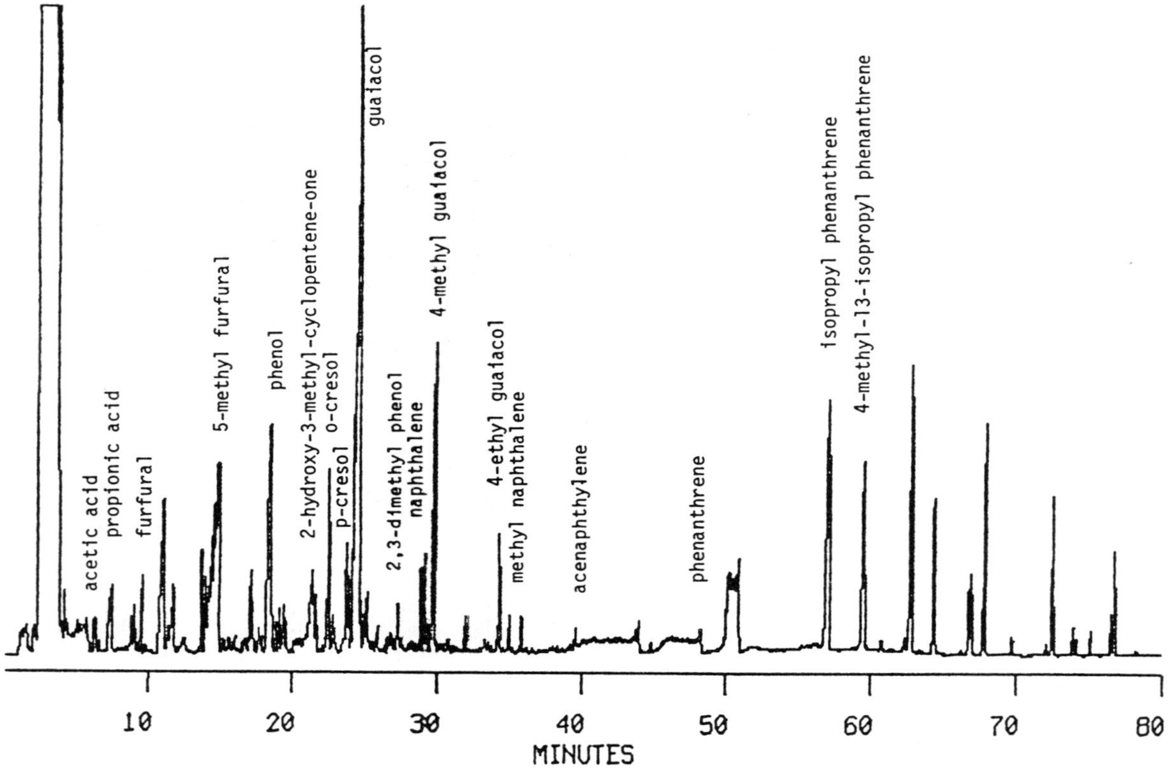

FIGURE 5. Volatiles of sugarcane bagasse tar (DB-5 cap. column, held 5 min. at 30°C, then to 280°C at 3°C/min).

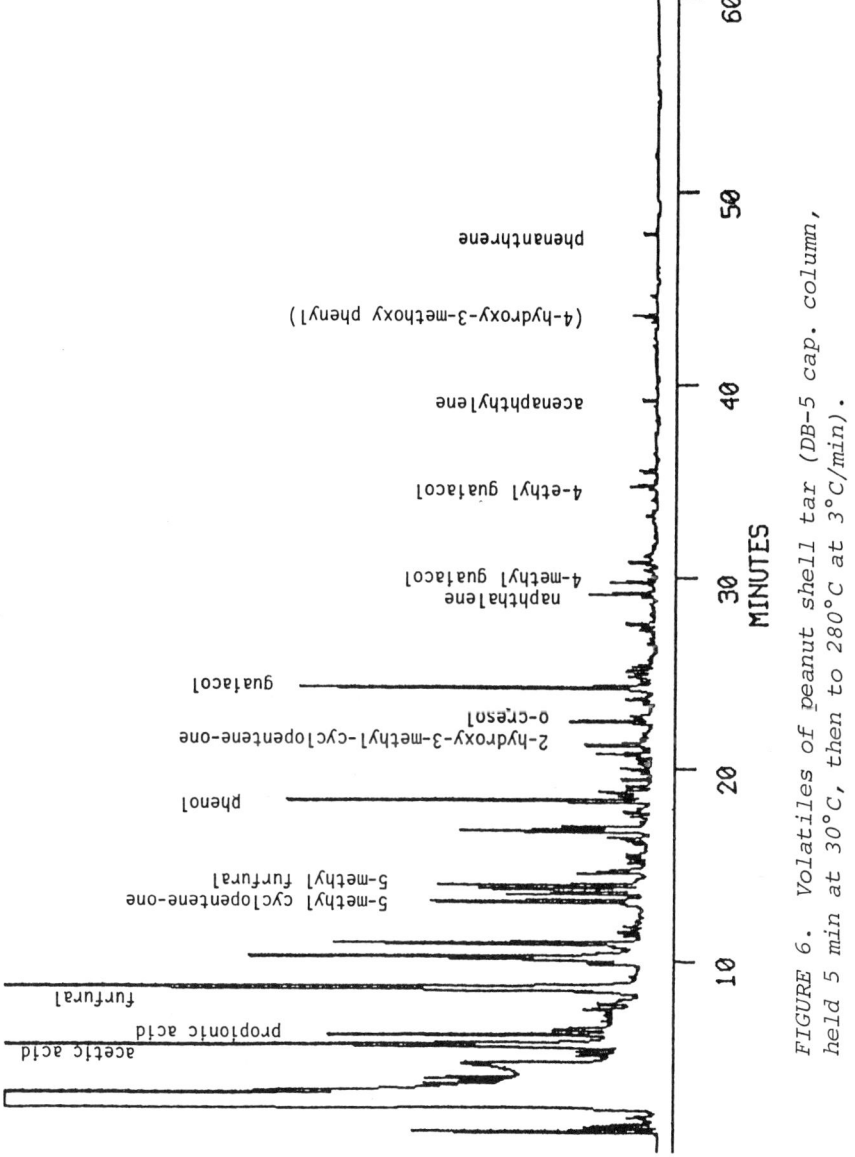

FIGURE 6. Volatiles of peanut shell tar (DB-5 cap. column, held 5 min at 30°C, then to 280°C at 3°C/min).

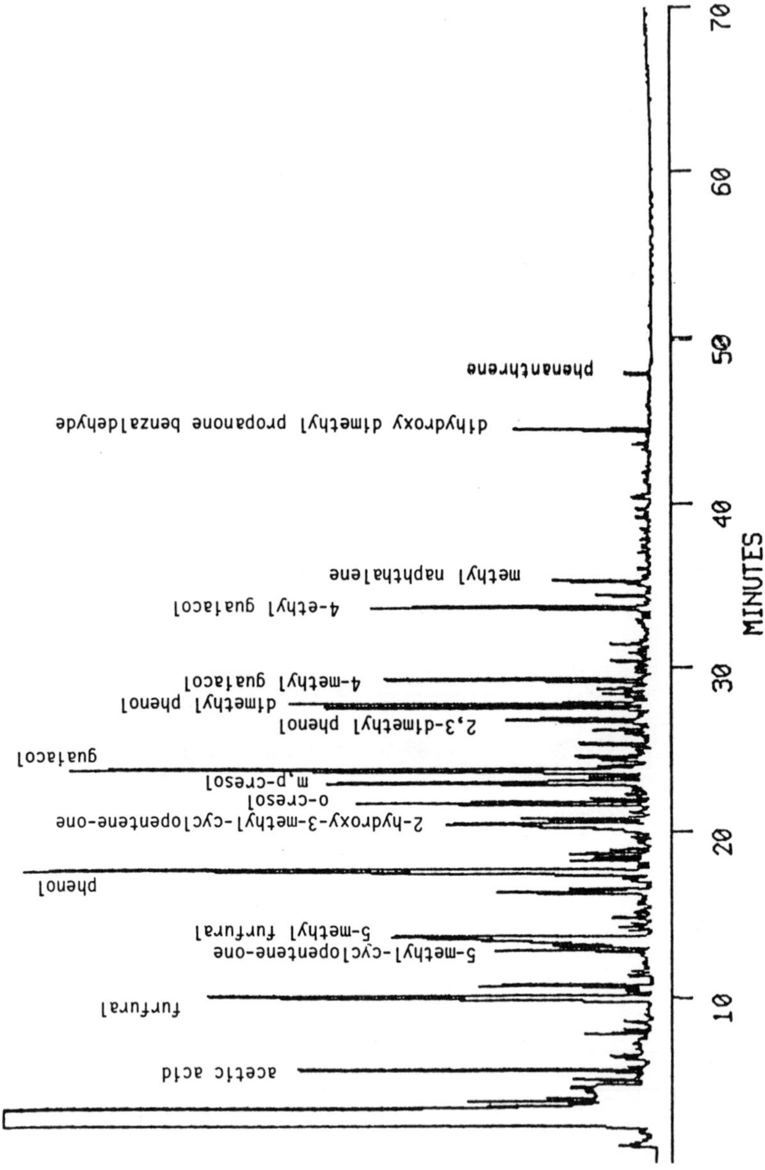

FIGURE 7. Volatiles of corncob tar (DB-5 cap. column, held 5 min at 30°C, then to 280°C at 3°C/min).

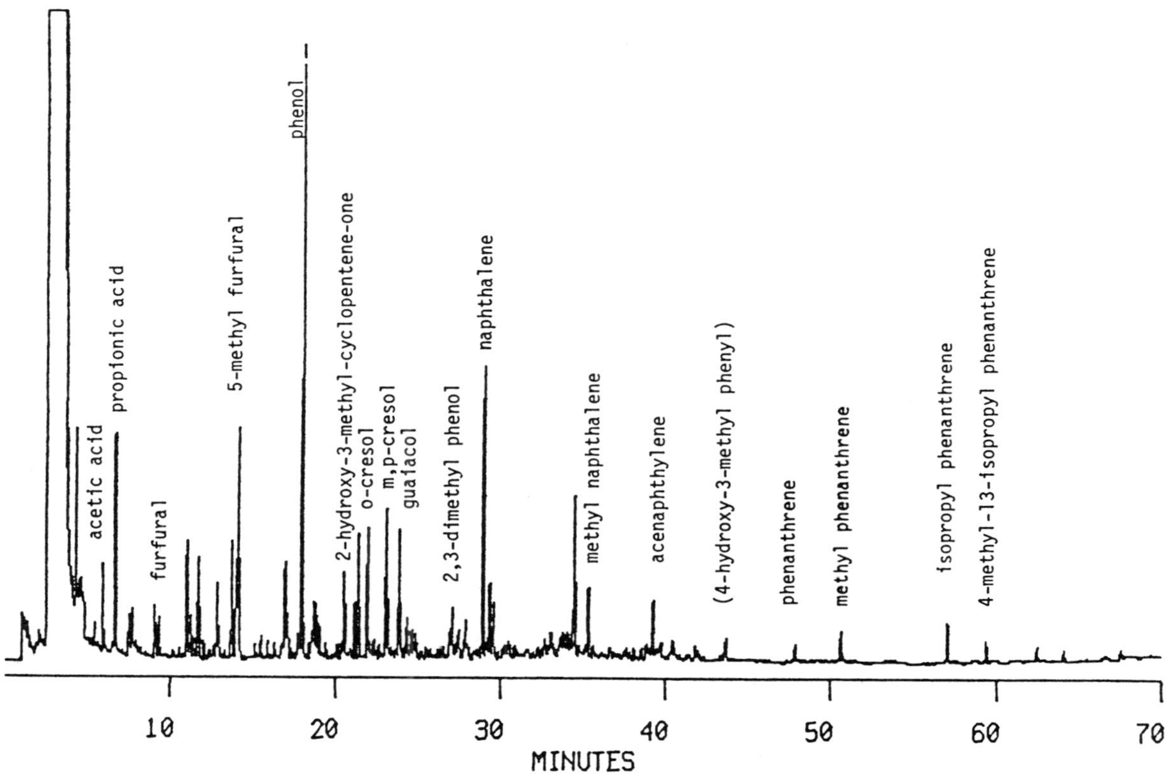

FIGURE 8. Volatiles of rice hull tar (DB-5 cap. column, held 5 min at 30°C, then to 280°C at 3°C/min).

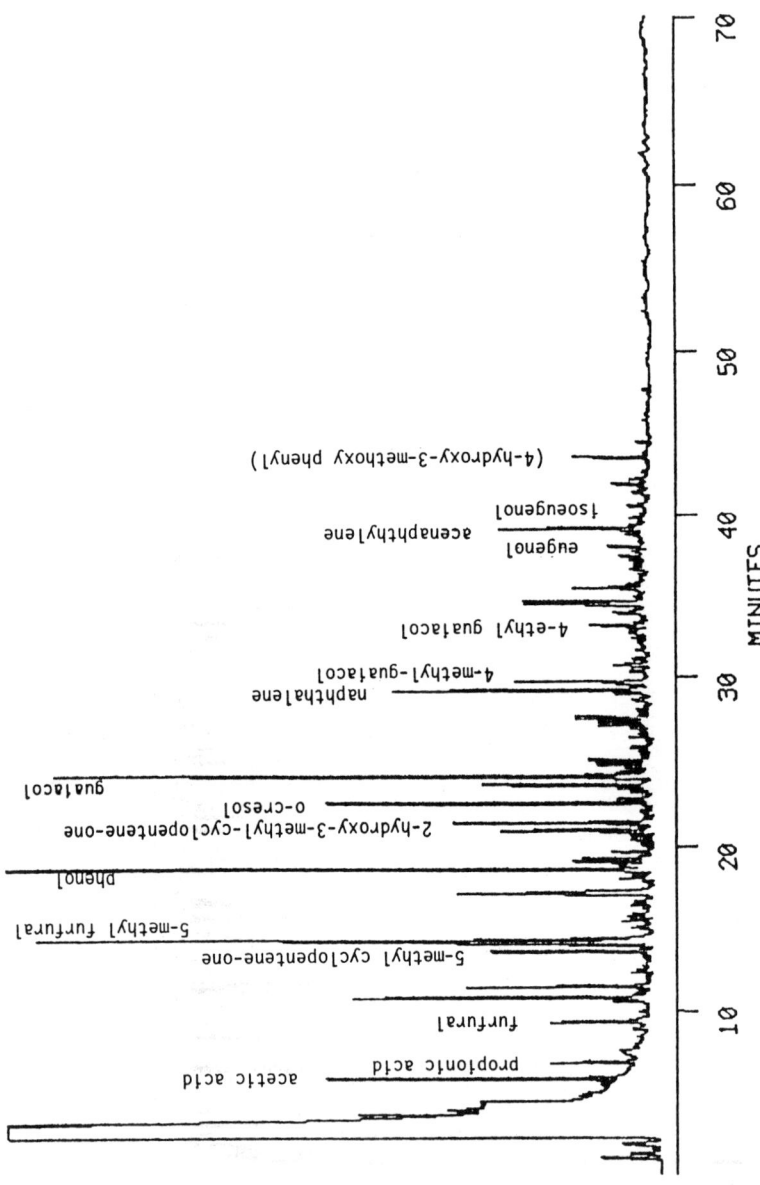

FIGURE 9. Volatiles of cottonseed hull tar (DB-5 cap. column, held 5 min at 30°C, then to 280°C at 3°C/min).

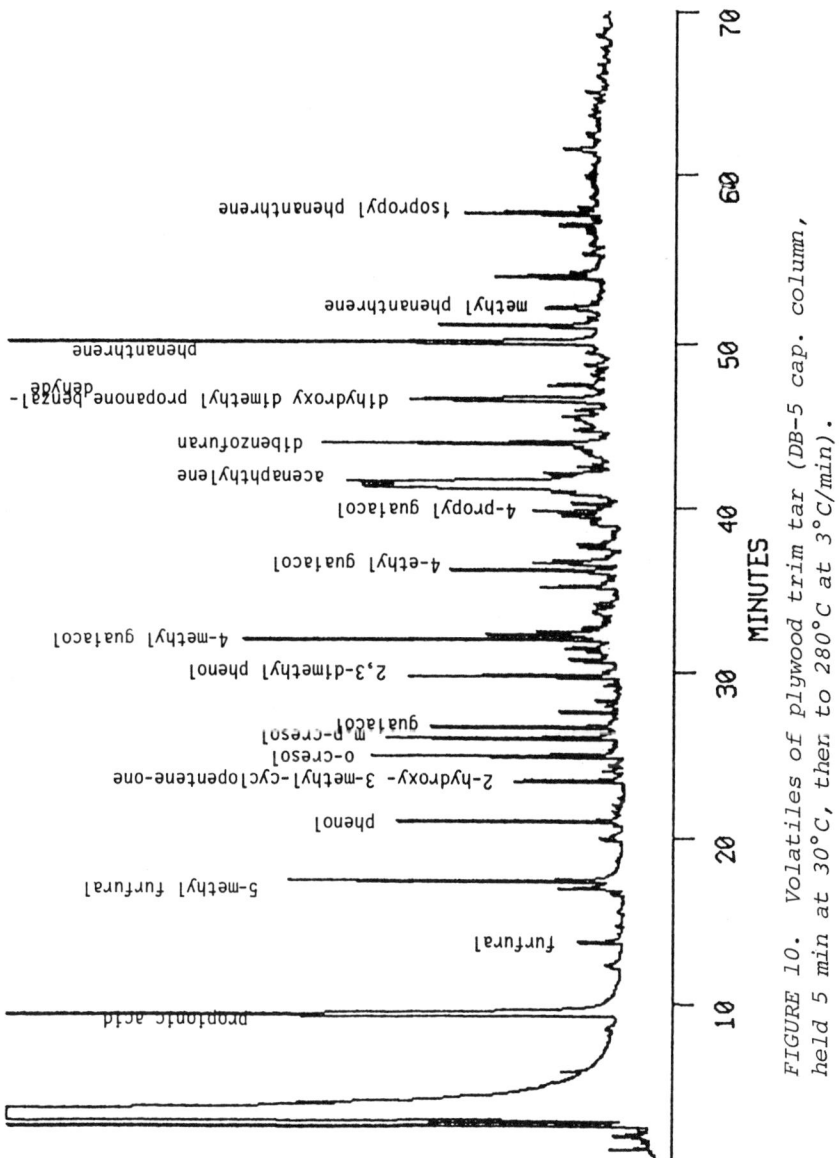

FIGURE 10. Volatiles of plywood trim tar (DB-5 cap. column, held 5 min at 30°C, then to 280°C at 3°C/min).

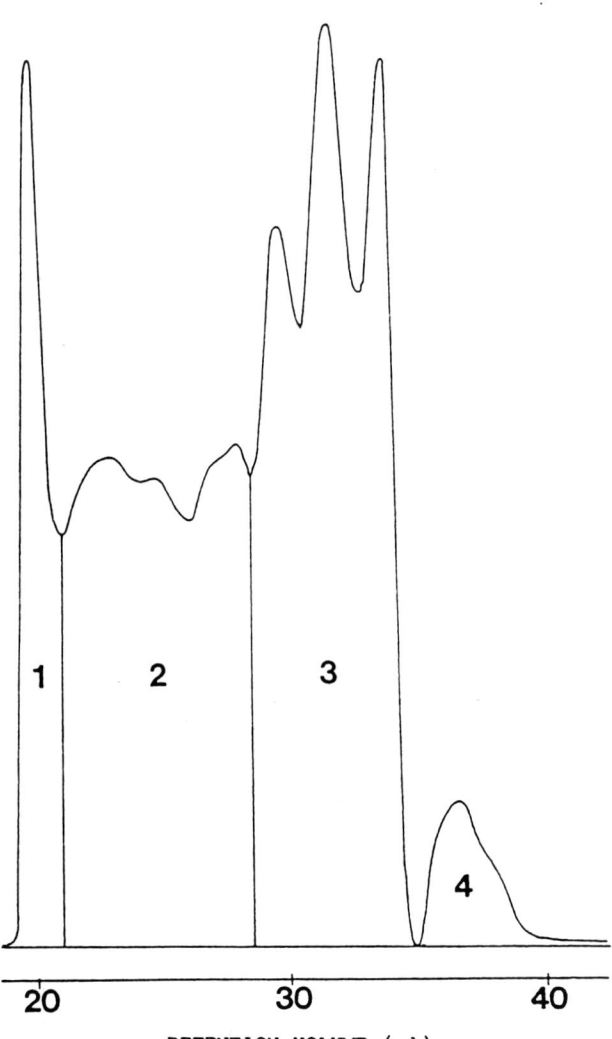

FIGURE 11. GPC of Tech-Air pyrolytic tar (Four 30 cm. 10nm micro-Styragel columns).

After the run, the specimens were cleaned, and precisely weighed. The mass loss during the test period was obtained. All corrosion products were removed from the specimens without removal of sound metal. This was accomplished by a bristle brush using a mild abrasive followed by cleaning with an appropriate solvent to remove the tar.

Types of corrosion found for each of the metals in the three solutions are given in Table V.

Table VI gives the weight loss data and the corrosion rates calculated using the following equation:

$$\text{Corrosion rate} = \frac{KW}{ATD}$$

where $K = 2.87 \times 10^2$

W = mass loss in grams

$A = 26.01 \text{ cm}^2$

$T = 166$ hrs, and

D = density of specimen in g/cm^2

The data shows that aluminum and steel are attacked severely by the three solutions. It was also observed that the prepared acid solution, which consists of the same composition of acetic and formic acid as the tar, is the most corrosive among the three treatments. Acetic acid together with formic acid in the pyrolysis tar are thus the major factors in tar corrosion. It may be that the corrosivity of tars in general, whether produced via biomass gasification or pyrolysis, is due to these acids in dilute concentrations. Although this may appear surprising, dilute solutions of acetic and formic acid have been reported to be corrosive to many steels (Biggs et al., 1961).

III. TAR HYDROPROCESSING

Although reported attempts at tar utilization usually relate to direct combustion, tar is a poor liquid boiler fuel (See Table VII). It is a still poorer internal combustion engine fuel. It is viscous, gummy, not completely volatile, corrosive, exhibits high oxygen contents and does not mix with conventional fuels. In order to use tar as an engine fuel, tar must be processed to reduce viscosity and gumming tendencies, improve volatility, remove acidity and lower oxygen content. As most of the undesirable properties are

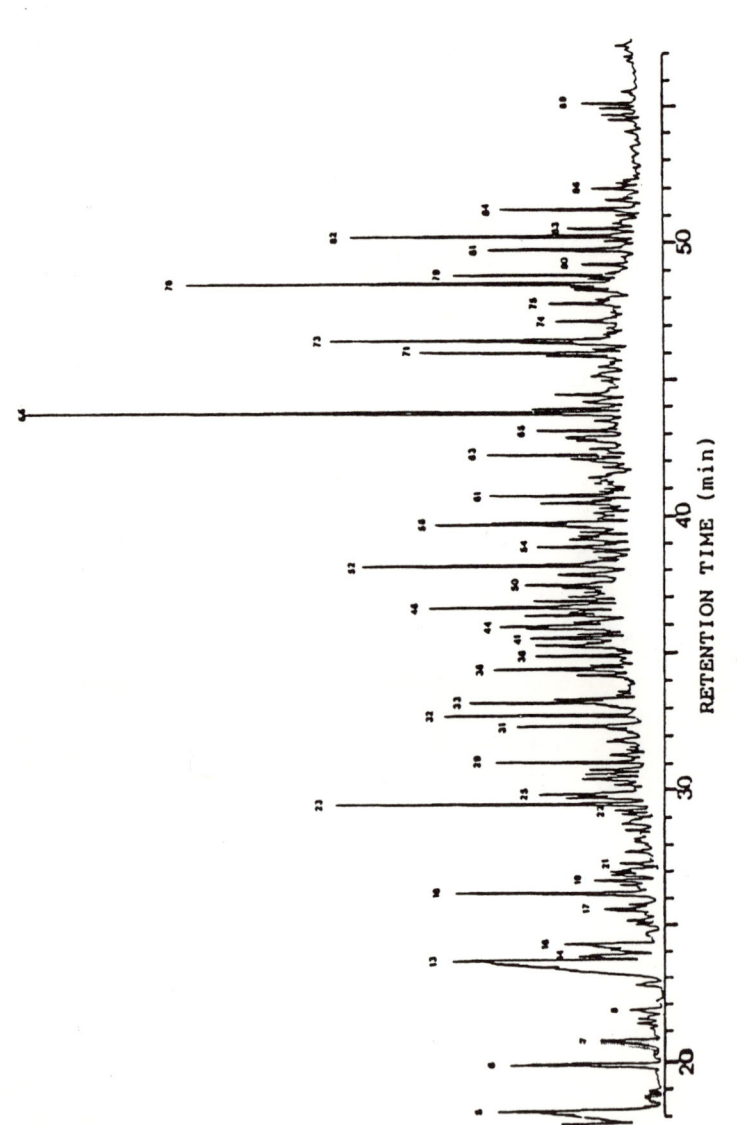

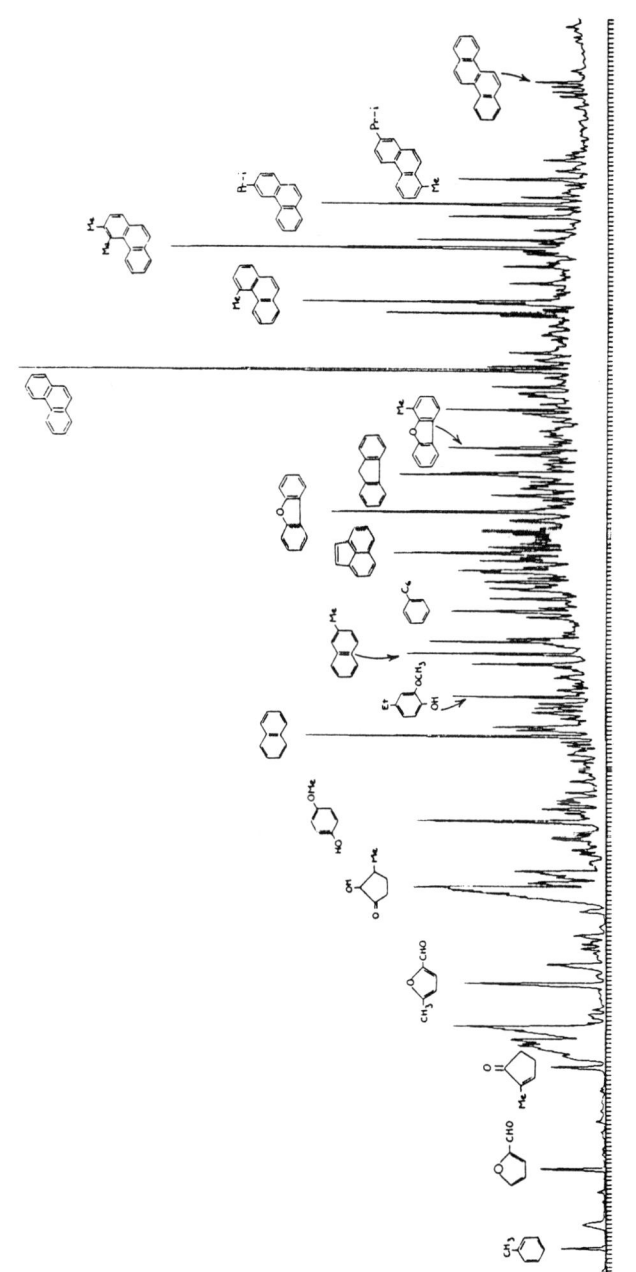

FIGURE 12. Aromatic fraction (No. 4) from GPC of Tech-Air pyrolytic tar (DB-5 cap. column, held 5 min at 30°C, then to 280°C at 4°C/min).

TABLE IV. Composition of the Aromatic Fraction in Band 4, GPC of Tech-Air Pyrolysis Tar

Peak number	Identification
1	toluene
2	2-furancarboxaldehyde
3	2-methyl-2-cyclopenten-1-one
4	1-(2-furanyl)-ethanone
5	$C_7H_{14}O$
6	methylfuraldehyde
7	3-furanone
8	dimethylfuranone
13	2-hydroxy-3-methyl-2-cyclopenten-1-one
14	3-isopropylcyclopentene
15	$C_7H_{16}O$
16	$C_7H_{16}O$
17	1-ethylcyclohexene
18	4-methoxyphenol (guaiacol)
19	5-methylbenzofuran
20	2-ethylbenzofuran
21	cyclohexylethanone
22	$C_8H_{14}O$
23	naphthalene
25	$C_8H_{16}O_2$
26	4,7-dimethylbenzofuran
27	$C_{10}H_{10}O_2$
30	2-methyl-3(2H)-benzofuran
31	2,3-dihydro-1H-inden-1-one
32	1-methylnaphthalene
33	2-methylnaphthalene
36	C6-alkyl benzene
37	C6-alkyl benzene
38	1,2-dihydro-acenaphthalene
43	dimethylnaphthalene
44	dimethylnaphthalene
45	dimethylnaphthalene
46	acenaphthalene
47	tetrahydro-dimethylnaphthalenone
48	tetrahydro-dimethylnaphthalenone

(Continued)

TABLE IV. *(Continued)*

Peak number	Identification
50	acenaphthene
52	dimethylbenzofuran
53	$C_{12}H_8O$
57	C_4-alkylnaphthalene
58	9H-fluorene
59	2-(1-methyl)-naphthalene
61	1-methyldibenzofuran
62	2-methyldibenzofuran
64	1-methoxyfluorene
65	2-methoxyfluorene
66	phenanthrene
67	anthracene
70	methylphenanthrene
71	methylphenanthrene
72	methylphenanthrene
73	methylphenanthrene
74	2-phenylnaphthalene
75	ethylphenanthrene
76	ethylphenanthrene
77	dimethylphenanthrene
78	dimethylphenanthrene
79	fluoranthene
80	pyrene
81	methylpyrene
82	isopropylphenanthrene
83	isopropylphenanthrene
84	C_4-alkylphenanthrene
85	trimethylphenanthrene
86	C_4-alkylphenanthrene
88	triphenylene
89	chrysene

TABLE V. Type of Corrosion in Metal Specimens Subjected to Pyrolysis Tar, Steam Distillate of Pyrolysis Tar and as Acid Solution Containing the Organic Acids Found in Pyrolysis Tar[a]

Solution	Aluminum	Brass	304 S.S.	316 S.S.	Steel	Copper
Pyrolysis tar	Concentration cell	Uniform attack	Pitting	Trace	Pitting	Uniform attack
Steam distillate	Concentration cell	Uniform attack	Trace	Trace	Pitting	Uniform attack
Acid solution	Uniform attack	Uniform attack	Trace	Trace	Pitting	Uniform attack

[a] Determination of corrosion type based on Bosich (1970).

TABLE VI. Corrosion Rate of Metals by Pyrolysis Tar, Steam Distillate of Pyrolysis Tar and an Acid Solution Containing the Organic Acids Found in Pyrolysis Tar

Test Parameter	Aluminum	Brass	304 S.S.	Steel	316 S.S.	Copper
Density (g/cc)	2.70	8.52	7.65	8.10	7.98	8.94
Pyrolysis tar						
Wt. loss, g.	0.0992	0.090	0.3734	0.3783	0.0011	0.0066
Corr. rate inch/min	2.442×10^{-3}	7.021×10^{-5}	3.247×10^{-3}	3.104×10^{-3}	9.162×10^{-6}	4.907×10^{-5}
Steam distillate						
Wt. loss, g.	0.0819	0.0036	0.0001	0.3142	0.0001	0.0013
Corr. rate inch/min	2.016×10^{-3}	2.808×10^{-5}	—	2.578×10^{-3}	—	9.655×10^{-6}
Acid solution						
Wt. loss, g.	0.2398	0.0276	0.0004	0.3344	0.0003	0.0245
Corr. rate inch/min	6.012×10^{-3}	2.153×10^{-6}	3.475×10^{-6}	2.627×10^{-2}	2.449×10^{-6}	1.821×10^{-4}

TABLE VII. Comparison of Fuel Oil Properties of No. 6 Fuel Oil and Pyrolysis Tar

	No. 6 fuel oil	Pyrolysis tar
C, wt %	85.7	65.8
H,	10.5	7.1
O,	2.0	27.1
S,	0.7-3.5	0.1
N,	-	0.1
Density, g/ml	0.98	1.14
Btu/lb	18200	9973[a]
Btu/gal	148800	94850[a]
Pour point, °C	18-29	32[a]
Flash point, °C	66	111[a]
Viscosity, SSU, 190°F	340	1150[a]
Pumping temp., °C	46	71[a]
Atomization temp., °C	105	116[a]

[a] *Pyrolysis tar containing 14% moisture.*

related to higher molecular weight components and hydrophilicity, catalytic hydrocracking and hydrotreating (hydroprocessing) were proposed.

A. Catalysts

Several types of catalysts were screened for hydrotreating and hydrocracking activity (see Table VIII). Initial work with 0.5% Pd and Pt catalysts on alumina with hydrogen-donor solvents such as tetralin and decalin gave promising results. However, the products were incompletely hydrotreated as shown by the high alkaline solubility, and examination by gas chromatography confirmed the presence of simple phenolics. Thus, these catalysts do not hydrotreat phenolics into aromatic hydrocarbons, at least not under the reaction conditions used.

Encouraged by the preliminary investigations with 0.5% Pt and Pd catalysts, a series of experiments was conducted with higher concentrations of noble metals on various supports. Use of these resulted in many improvements. There was more hydrogen uptake in the production of water (more oxygen

TABLE VIII. Catalysts Screened for Hydrotreating and Hydrocracking of Biomass Thermochemical Tars

5% Pd/alumina	0.5% Pd/alumina
5% Pd/carbon	0.5% Pt/alumina
	0.5% Re/alumina
5% Pt/alumina	
5% Pt/carbon	Harshaw Ni-4301
	Harshaw CoMo-0603
5% Re/alumina	Harshaw HT-400
5% Rh/alumina	
5% Ru/alumina	Silica alumina
	$NiO-WO_3$ on silica alumina
Raney Ni	
$NiCO_3$	
UOP Lomax	ZrO_2 on alumina
UOP Unibon	

removal), in the breakage of C-O bonds (removal of phenolic hydroxyl and methoxyl substituents, some cracking and reduction of acidity), in the removal of unsaturation (leading to products of light color), and in hydrocracking (higher yields of liquid product with lower viscosities). A number of these catalysts produced essentially the same results (Soltes, 1983d). The 5% Platinum and Palladium on alumina and activated carbon catalysts were subsequently used in this research, except as otherwise noted.

B. *Hydrogen-Donor Solvents*

Tetralin and decalin work well as hydrogen-donor solvents, but exhibit boiling points in the diesel fuel range. Thus, their use prevents both their recovery and the evaluation of the hydroprocessed product, or a distillative fraction thereof, as diesel fuel (Soltes and Lin, 1983a). More recently, the light ends of the hydrotreated product have been evaluated as solvent. The light ends contain the alkyl cyclohexanes, especially methyl cyclohexane, which appear to function just as well as decalin in hydrogen-donor activity in the system. These alkyl cyclohexanes are produced from the alkyl quaiacols present in the tar. Under the catalytic reaction conditions used, the phenolic hydroxyl and methoxyl moities of pine pyrolysis tar phenolics are split off the ring, followed by ring saturation to produce the alkyl cyclohexanes.

C. Hydroprocessing Conditions

Tars were hydroprocessed in 1.5 L batch rocking reactors rated at 5000 psi and 500°C (Aminco 11.1 cm o.d. Series). With initial hydrogen pressure set at 6900 kPa cold, and temperature elevated to and maintained at 400°C for 60 minutes, hydrogen saturation is achieved. Results are given in Table IX for hydroprocessing of the tars under these conditions in methyl cyclohexane hydrogen-donor solvent. Decalin/tetralin solvent results are given elsewhere (Soltes, 1983c; Soltes and Lin, 1983a).

IV. COMPOSITION AND UTILITY OF HYDROPROCESSED TARS

Some chemical and physical characteristics of the hydroprocessed tars are given in Table X. Note that the values given are not corrected for moisture content. Typically, hydroprocessed tars can contain a few percent water.

A. Pine Waste Pyrolysis Tar and Corn Cob Gasification Tar

Hydroprocessed Tech-Air pyrolysis tar was subjected to several GC-mass spectroscopy runs. The mass spectra generated in these runs resulted in the identification of nearly 200 components (see Table XI; Soltes, 1983c). These were used to label the peaks in Figures 13 and 14: the Tech-Air pine waste and DeKalb corncob hydroprocessed tars using methyl cyclohexane as solvent, followed by solvent stripping.

B. Other Hydroprocessed Tars

Additional GC-mass spectrometry on hydroprocessed tars or their fractions were used to label the chromatograms of other hydroprocessed tars which follow (Figures 15 through 22). Note the similarities in the composition of these hydroprocessed tars. It should be expected that tar hydroprocessing using catalysts selected for hydrogenolysis activity for carbon-oxygen bond cleavage would eradicate any differences in methoxyl substitution, and further remove phenolic hydroxyl to yield alkyl aromatics and, under saturation conditions, alkyl cyclohexanes. We can report that the volatiles of the hydroprocessed tar products are indeed similar in composition. Compositions of these hydroprocessed tars reflect not only the alkyl aromatics expected, but also a mixture of paraffinic hydrocarbons.

TABLE IX. Hydroprocessing of Biomass Tars in Methyl Cyclohexane

Tar used	Wt. Tar (wet) g.	Oil + Solv. g.	Water g.	Water (%)	Conversion (wt%)	Conversion Energy %	Gas + loss (%)
Tech Air	67.2	137.0	17.0	(34.2)	(61.2)	95.5	(4.6)
Wood chips	69.0	131.0	16.0	(23.2)	(48.0)	70.5	(28.8)
Pecan shell	65.5	131.0	20.0	(30.5)	(54.4)	79.6	(15.1)
Bagasse	60.0	130.3	15.5	(25.8)	(62.8)	86.0	(11.4)
Peanut shell	68.0	136.8	22.0	(32.4)	(63.7)	75.0	(3.9)
Corncob	57.0	126.8	10.2	(17.9)	(48.5)	74.3	(33.6)
Rice hull	64.3	138.7	6.7	(10.4)	(63.4)	92.7	(26.2)
Cotton hull	50.9	125.1	10.5	(20.6)	(51.9)	84.8	(27.5)
Plywood trim	74.0	140.2	18.0	(24.3)	(68.4)	96.7	(7.3)

Conditions used: 15g. 5% Pd on Alumina Catalyst; 100 g. methyl cyclohexane; 400°C, 60 minutes; 6900 kPa hydrogen.

$$\text{Conversion, wt.\%} = \frac{(100)(Oil+Solv - Solv)}{Tar\ dry\ wt.}, \text{ moisture from Table I}$$

$$\text{Conv., energy\%} = \text{conversion, wt.\%} \times \frac{\text{heat content, processed tar}}{\text{heat content, tar}}$$

Gas + Loss, % = 100 - % water - % conversion.

TABLE X. Chemical and Physical Characteristics of Hydroprocessed Biomass Tars

	Tech-Air	Wood chips	Pecan shell	Sugar-cane bagasse	Peanut shell	Corn cob	Rice hull	Cotton hull	Plywood trim
Elemental analysis[a]									
C, wt. %	83.20	79.48	81.58	78.62	78.50	83.40	81.78	82.05	81.14
H,	12.70	9.69	10.96	9.28	9.85	9.22	10.53	10.80	9.51
O,	4.10	10.57	6.41	11.25	10.24	5.49	5.53	5.53	8.45
N,	–	0.26	1.06	0.75	1.41	1.89	2.16	1.38	0.90
Heat content									
MJ/kg[a]	40.26	38.19	40.24	38.15	38.33	39.32	37.37	41.02	38.91
Specific gravity[a]	0.87	0.90	0.90	0.94	0.95	0.88	0.93	0.90	0.89

[a] Not corrected for any moisture present in the product.

TABLE XI. *Identified Chemical Components in Hydroprocessed Tech-Air Pine Waste Pyrolysis Tar*

Identifications by GC-Mass Spec.
Listed in order of elution on DB-5 column

pentane
cyclopentane
hexane
methyl cyclopentane
cyclohexane

3-hexanone
methyl cyclopentene
2-cyclohexen-1-ol
1,3-dimethyl cyclopentane
tetrahydro-2H-pyran

heptane
methyl cyclohexane
ethyl cyclopentane
tetrahydro-2-methyl-2H-pyran
1,2,4-trimethyl cyclopentane

methyl benzene
2-ethyl tetrahydrofuran
2-methyl heptane
isopropyl cyclopentane
trans-1,3-dimethyl cyclohexane

trans-1-ethyl-3-methyl cyclopentane
2-propenyl cyclohexane
trans-1,2-dimethyl cyclohexane
cis-1,4-dimethyl cyclohexane
octane

cis-1,3-dimethyl cyclohexane
cis-1-ethyl-3-methyl cyclopentane
trans-1-methyl-2-ethyl cyclopentane
cis-1,2-dimethyl cyclohexane
ethyl cyclohexane

cis-1-methyl-2-ethyl cyclopentane
1,3,5-trimethyl cyclohexane
3-methyl-7-oxabicyclo-[4,1,0]-heptane
trans-1-methyl-2-ethyl cyclohexane
ethyl benzene

(Continued)

TABLE XI. *(Continued)*

1,2,3-trimethyl-[1-alpha,2-alpha,3-beta] cyclohexane
isononane
1,3-dimethyl benzene
1-undecyne
2-methyl octane

2-ethyl-1,1-dimethyl cyclopentane
1-ethyl-3-methyl cyclohexane
1,2-dimethyl benzene
3,4,4-trimethyl-2-hexene
1-ethyl-2-methyl cyclohexane

1,2,3-trimethyl cyclohexane
1-ethyl-4-methyl cyclohexane
nonane
1-methylethyl cyclohexane
1-ethyl-1-methyl cyclohexane

propyl cyclohexane
7-methyl-1-octene
2,3-dimethyl octane
3-methyl-6-(1-methylethyl) cyclohexane
1,3-diethyl cyclohexane

propyl benzene
trans-octahydro-1H-indene
1-ethyl-2-methyl benzene
1-ethylmethyl benzene
1-ethyl-3-methyl benzene

3,7-dimethyl-6-octen-1-ol
3-decyne
[1S,3S]-(+)-M-methane
1-decene
cis-octahydro-1H-indene

(2-methylpropyl)-cyclohexane
trans-1-methyl-4-(1-methylethyl) cyclohexane
1,3,5-trimethyl benzene
cis-1-methyl-4-(1-methylethyl) cyclohexane
phenol

1-ethyl-4-methyl benzene
1,2-diethyl cyclohexane
4-ethyl-3-octene
2-methylpropyl-cyclohexane
decane

(Continued)

TABLE XI. (Continued)

1-ethenyl-2-methyl benzene
isobutyl benzene
1-methyl-4-propyl benzene
1,3-diethyl benzene
1,3-dimethyl-5-ethyl benzene

butyl cyclohexane
1-methyl-2-propyl benzene
trans-decahydro naphthalene
o-cresol

butyl benzene
2,3-dihydro-1-methyl-1H-indene
p-cresol
cis-decahydronaphthalene
undecane

diethyl toluene
2-pentyl-2-methyl butane
trimethyl ethyl benzene
1-pentyl-2-methyl butane
pentyl cyclohexane

2,4-dimethyl phenol
pentyl benzene
2-indanone
2,6-dimethyl phenol
1-ethenyl-3-ethyl benzene

4,6-decadiene
2-methyl decahydronaphthalene
2,3-dihydro-5-methyl-1H-indene
tetrahydronaphthalene
1,2-dimenthylpropyl benzene

(1,1-dimethyl-2-propenyl)-benzene
1,11-dodecadiene
naphthalene
dimethyl indane
3,4-dimethyl phenol

2-methyl-1,2,3,4-tetrahydronaphthalene
dodecane
butyl phenol
3,4-dihydro-2-naphthalenone
2-ethyl-4-methyl phenol

(Continued)

TABLE XI. (Continued)

1-methyl-1,2,3,4-tetrahydronaphthalene
m-isopropyl phenol
iso-dodecane
6-methyl-1,2,3,4-tetrahydronaphthalene
2-ethyl-6-methyl phenol

5-methyl-1,2,3,4-tetrahydronaphthalene
cyclopentylmethyl cyclohexane
cyclohexyl benzene
1,1-ethylidene-bis-cyclopentane
6-ethyl-1,2,3,4-tetrahydronaphthalene

alpha-ethyl-benzene methanol
2-methyl-trans-1,1-bicyclohexyl
5-methyl-1,2,3,4-tetrahydronaphthalene
3-ethyl-5-methyl phenol

2-methyl-cis-1,1-bicyclohexyl
1,1-methylene-bis-cyclohexane
tridecane
1,2,3,4-tetrahydro-1,5,7-trimethyl naphthalene
phenyl cyclohexanol

methyl methoxy phenyl acetate
isotridecane
methyl bytyl phenol
para-n-amyl phenol
methyl cyclohexyl cyclohexane

tetradecane
1,1-(1,2-ethanediyl)bis-cyclohexane
octahydro-2,3A,4-trimethyl-2-(1-methylethyl)-
  1H-indene
1-cyclohexyl octane

1,12-tridecadiene
1,1-(1,2-ethanediyl)-bis-cyclohexane
cyclohexyl-ethyl benzene
pentadecane
1,1-(1-methyl-1,2-ethanediyl)-bis-cyclohexane

7-(1,1-dimethylethyl)-3,4-dihydro-1-(2H)-
  naphthalenone
1,2-dicyclohexyl ethane
1-phenyl-2-cyclohexyl ethane
2,6,10-trimethyl dodecane

(Continued)

TABLE XI. (Continued)

---

hexadecane
2,4-dimethyl biphenyl
1-benzylidene-2,2,3-trimethyl cyclopentane
4-phenyl benzaldehyde
0,1-1,2-dimethyl-1,2-ethandiyl)-bis-cyclohexane

octahydrophenanthrene
heptadecane
phenyl undecane
phenanthrene
1,1-(1,5-pentanediyl)-bis-cyclohexane

octadecane
nonadecane
3-methyl phenanthrene
eicosane
1-phenyl-4-cyclohexyl-2-methyl cyclohexane

2,3-dimethyl phenanthrene
10,13-octadeca-diynoic acid methyl ester
heneicosane
1-methyl-7-(1-methylethyl) phenanthrene

2,3,5-trimethyl phenanthrene
docosane
2-butyl-8-hexyl naphthalene
2,6-bibutyl-4-ethyl phenol
tricosane

1,2,3,4,4A,9,10, 10A-octahydro-1,4A-dimethyl-1-
    phenanthrene carboxaldehyde
evodionol
alloevodionol
dehydroabietate

tetracosane
pentacosane
hexacosane
heptacosane
octacosane

nonacosane
triacontane

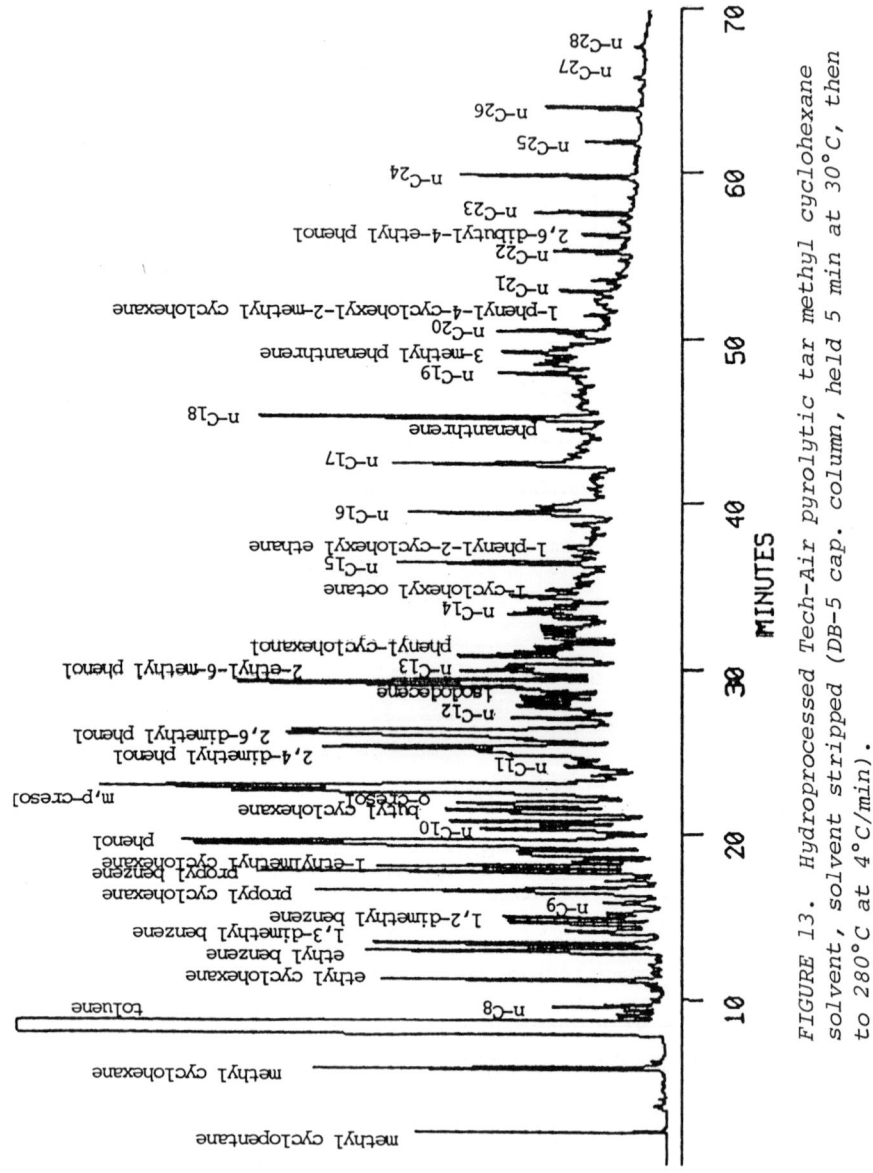

FIGURE 13. Hydroprocessed Tech-Air pyrolytic tar methyl cyclohexane solvent, solvent stripped (DB-5 cap. column, held 5 min at 30°C, then to 280°C at 4°C/min).

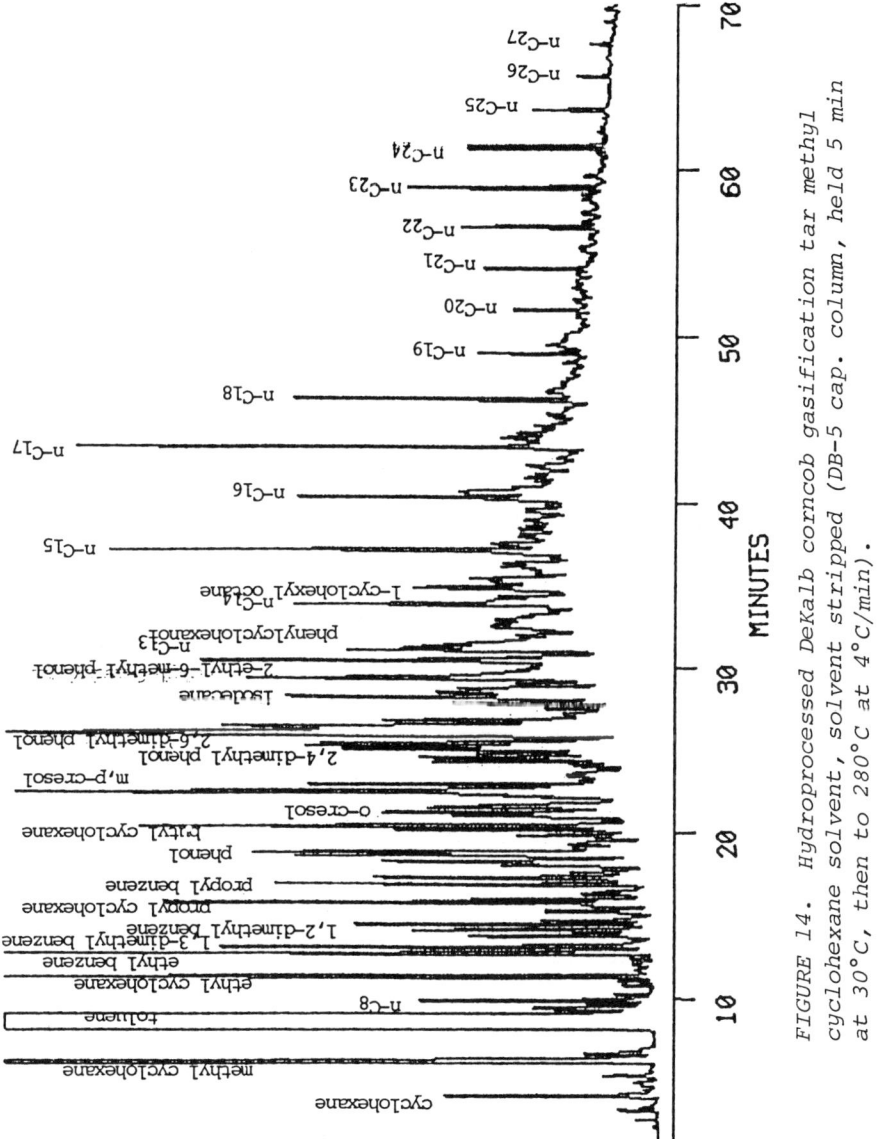

FIGURE 14. Hydroprocessed DeKalb corncob gasification tar methyl cyclohexane solvent, solvent stripped (DB-5 cap. column, held 5 min at 30°C, then to 280°C at 4°C/min).

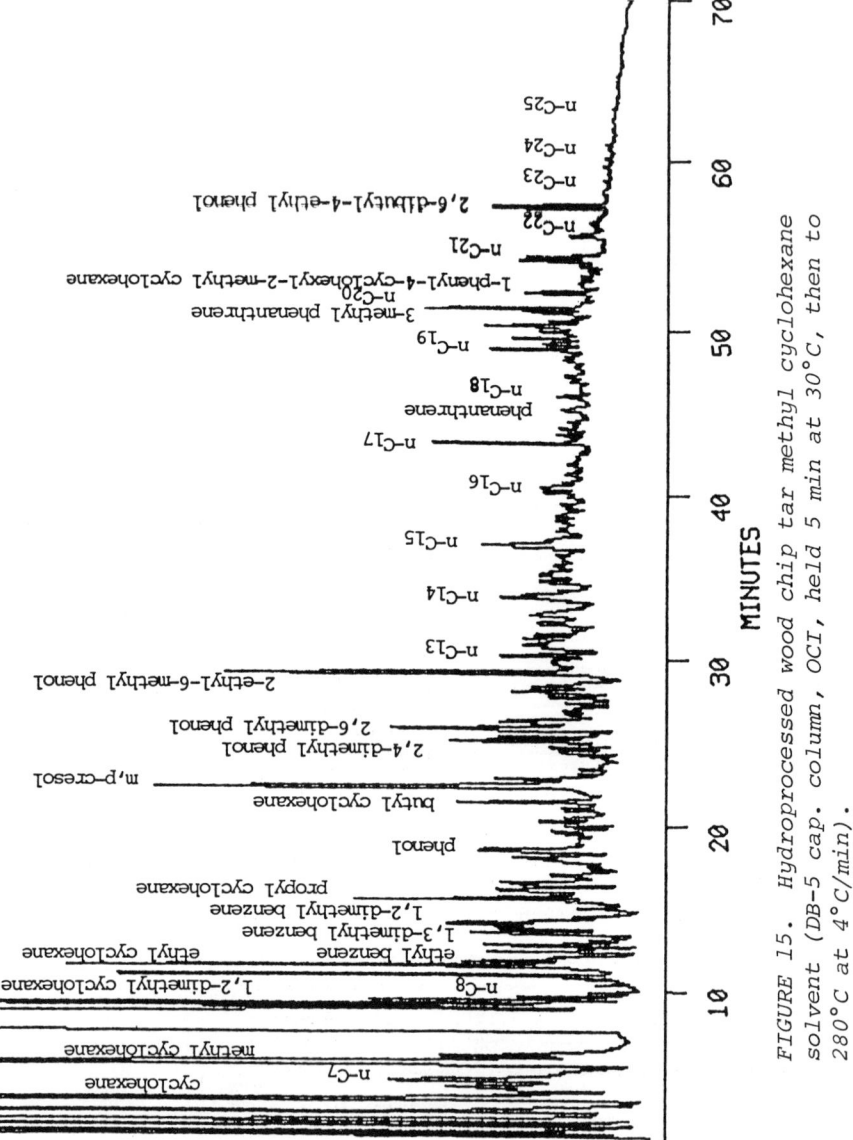

FIGURE 15. Hydroprocessed wood chip tar methyl cyclohexane solvent (DB-5 cap. column, OCI, held 5 min at 30°C, then to 280°C at 4°C/min).

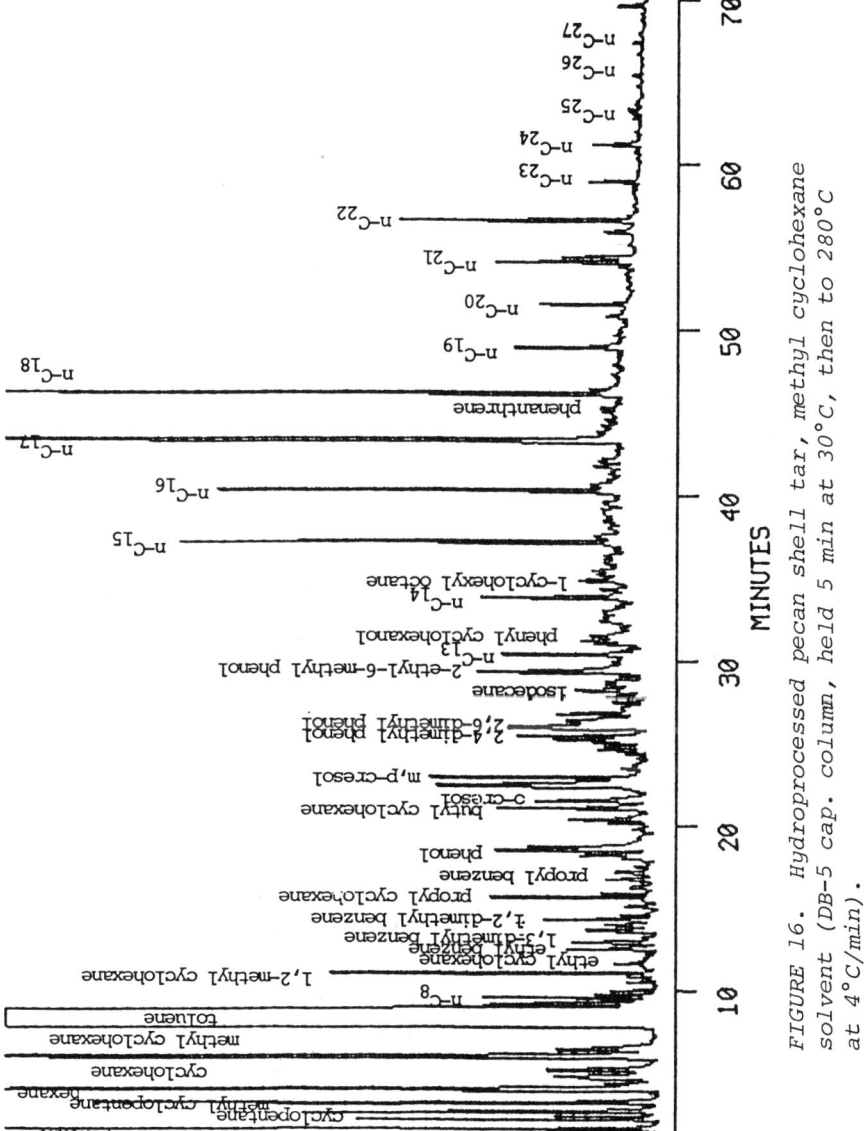

FIGURE 16. Hydroprocessed pecan shell tar, methyl cyclohexane solvent (DB-5 cap. column, held 5 min at 30°C, then to 280°C at 4°C/min).

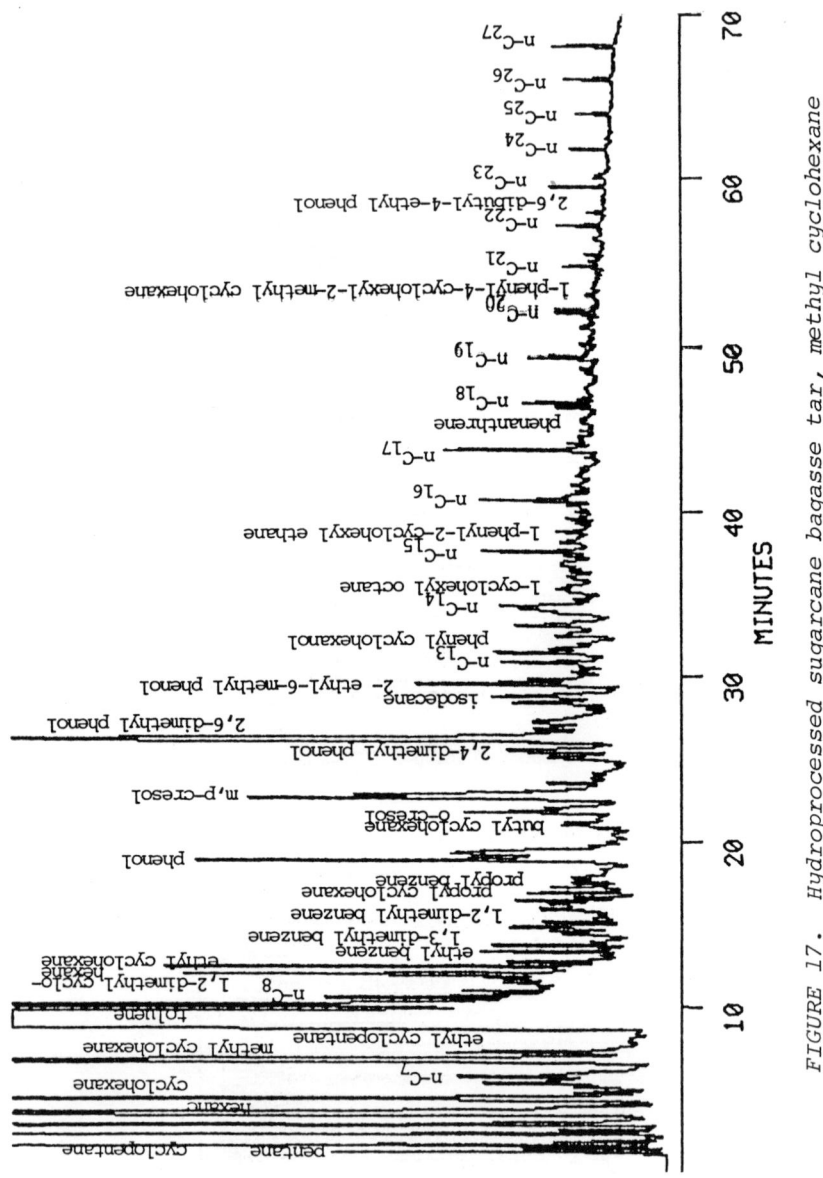

FIGURE 17. Hydroprocessed sugarcane bagasse tar, methyl cyclohexane solvent (DB-5 cap. column, held 5 min at 30°C, then to 280°C, at 4°C/min).

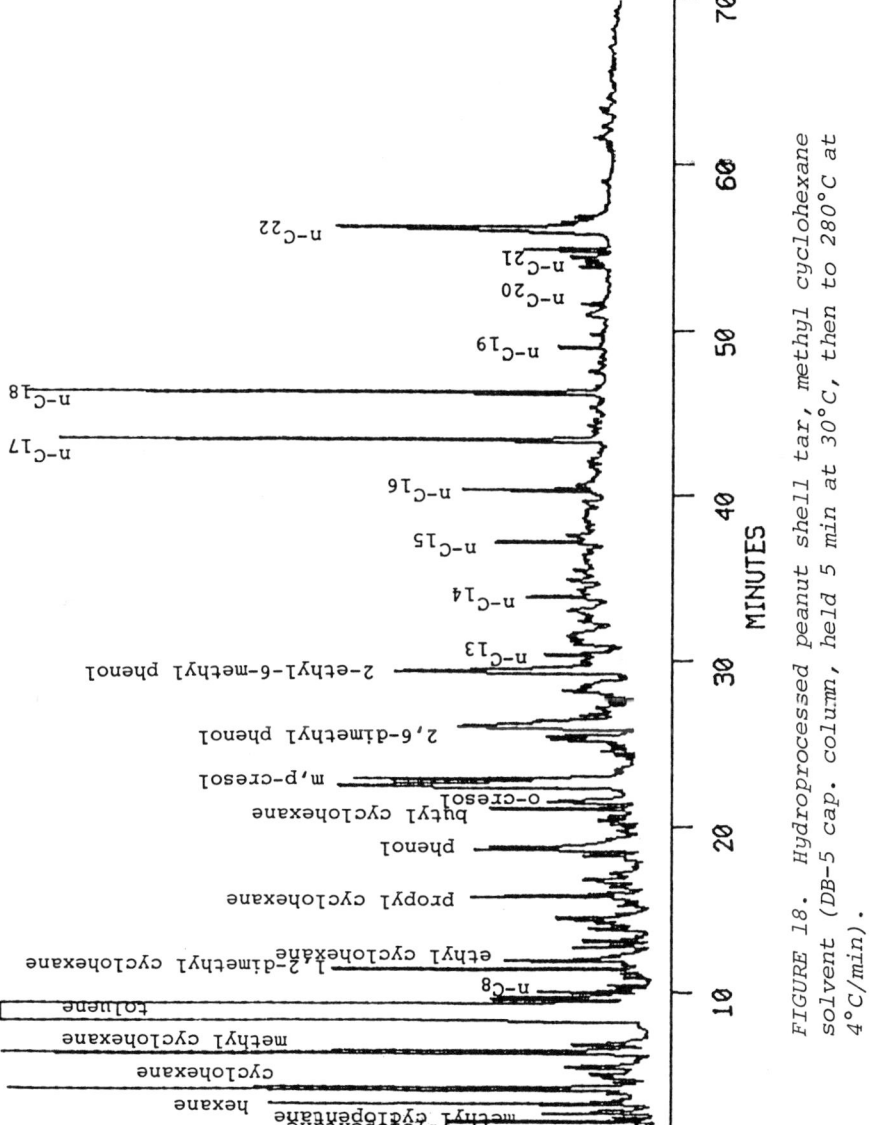

FIGURE 18. Hydroprocessed peanut shell tar, methyl cyclohexane solvent (DB-5 cap. column, held 5 min at 30°C, then to 280°C at 4°C/min).

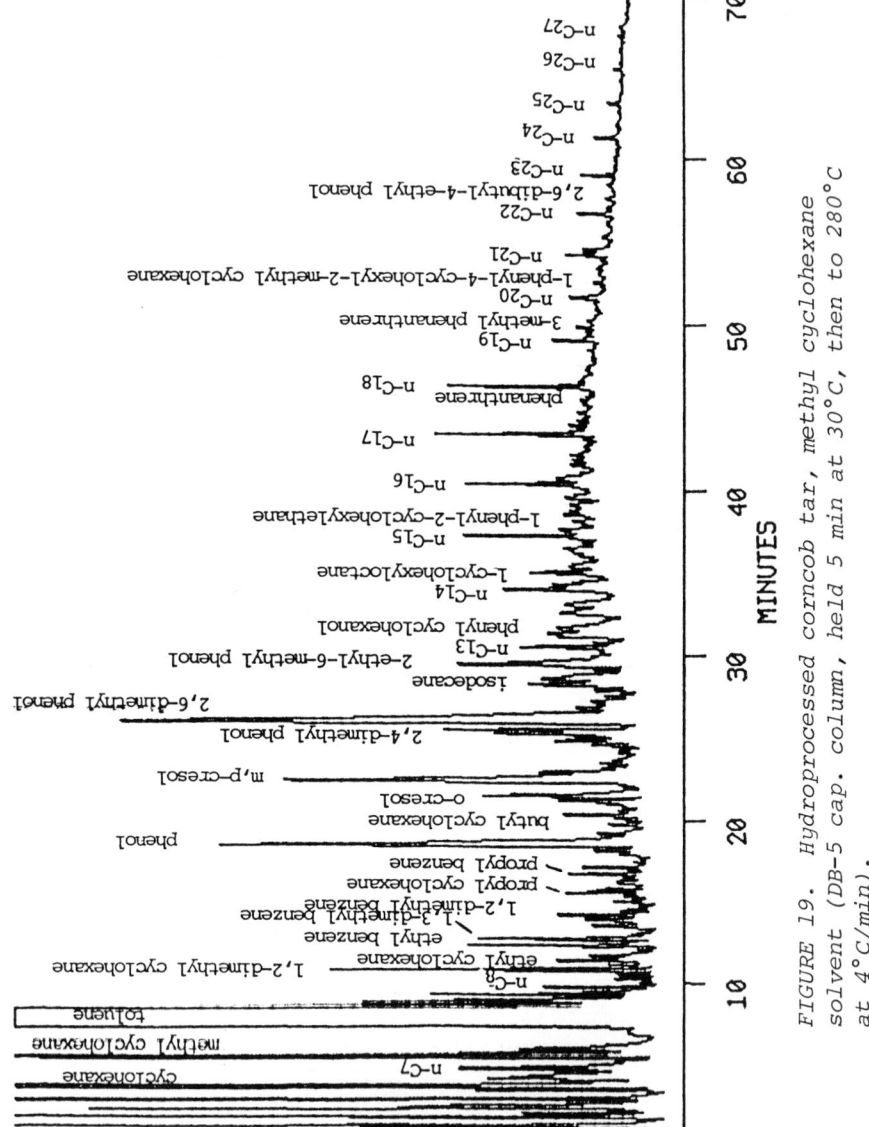

FIGURE 19. Hydroprocessed corncob tar, methyl cyclohexane solvent (DB-5 cap. column, held 5 min at 30°C, then to 280°C at 4°C/min).

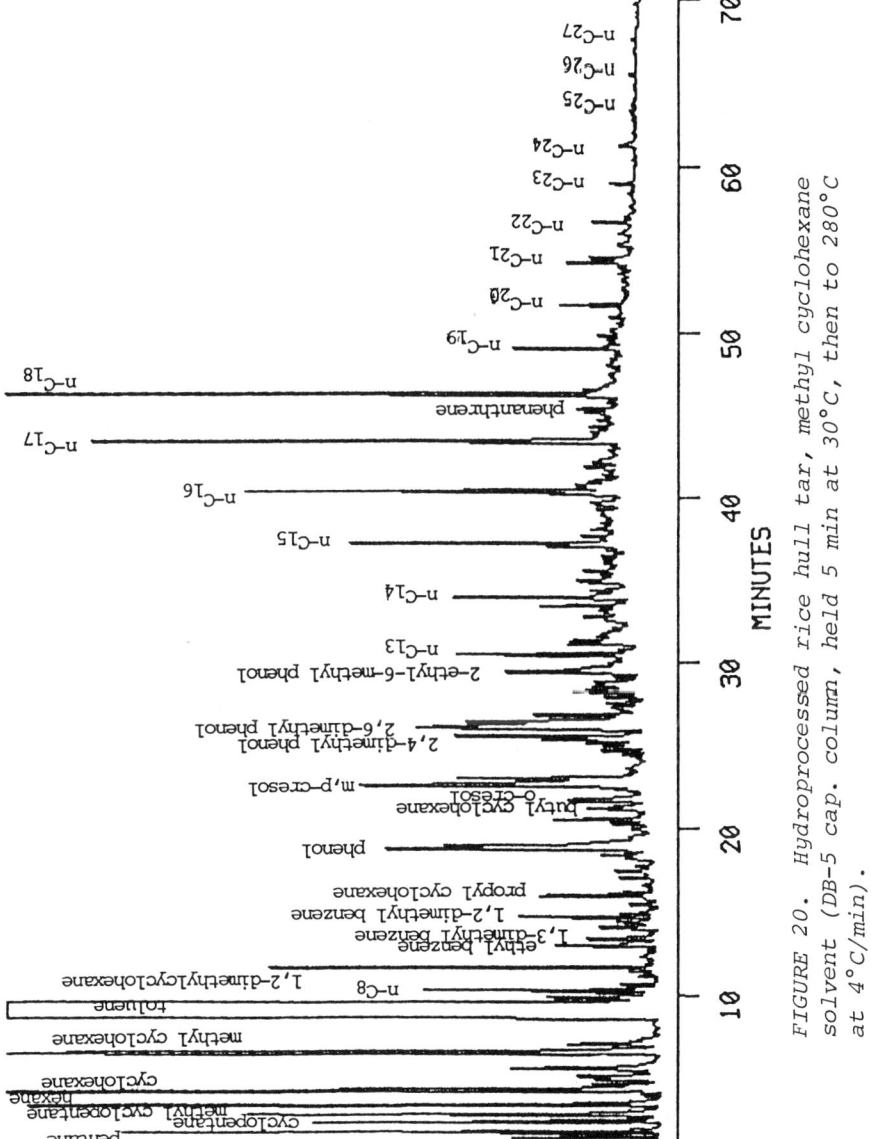

FIGURE 20. Hydroprocessed rice hull tar, methyl cyclohexane solvent (DB-5 cap. column, held 5 min at 30°C, then to 280°C at 4°C/min).

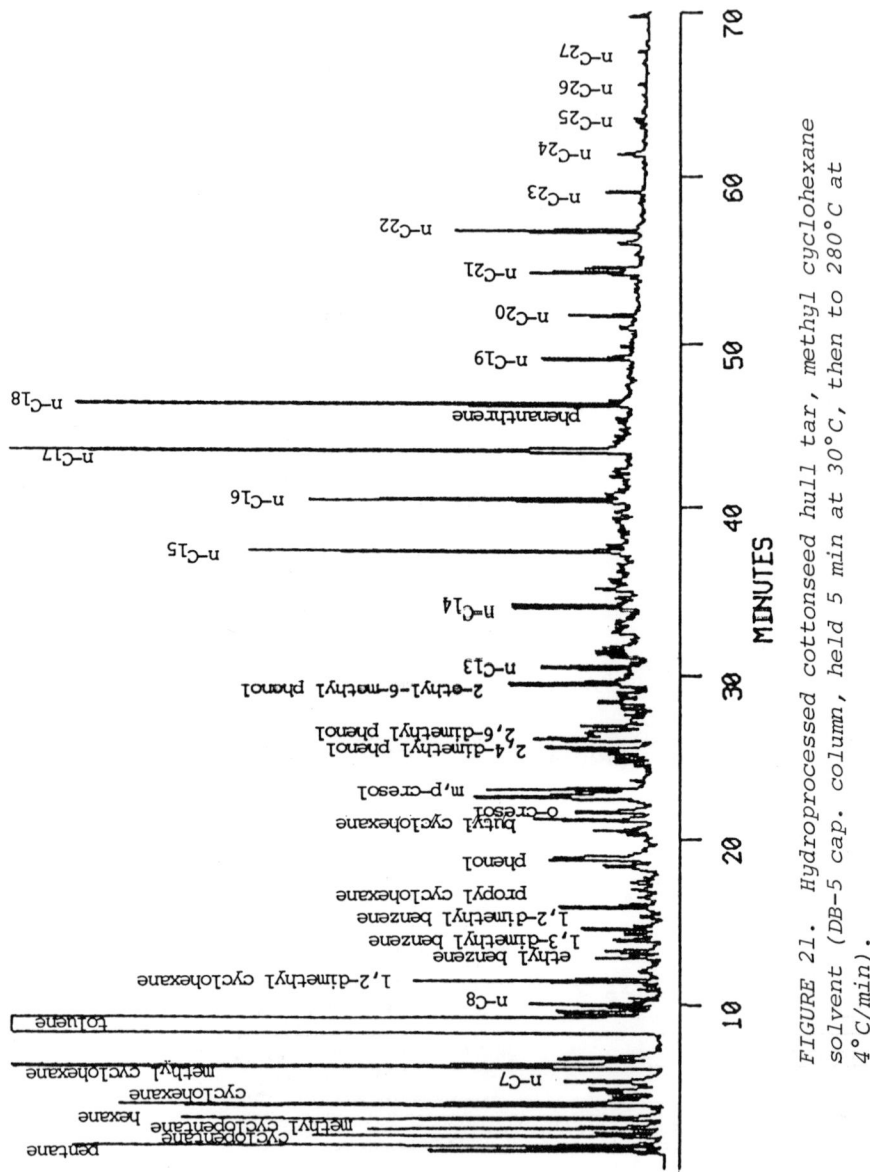

FIGURE 21. Hydroprocessed cottonseed hull tar, methyl cyclohexane solvent (DB-5 cap. column, held 5 min at 30°C, then to 280°C at 4°C/min).

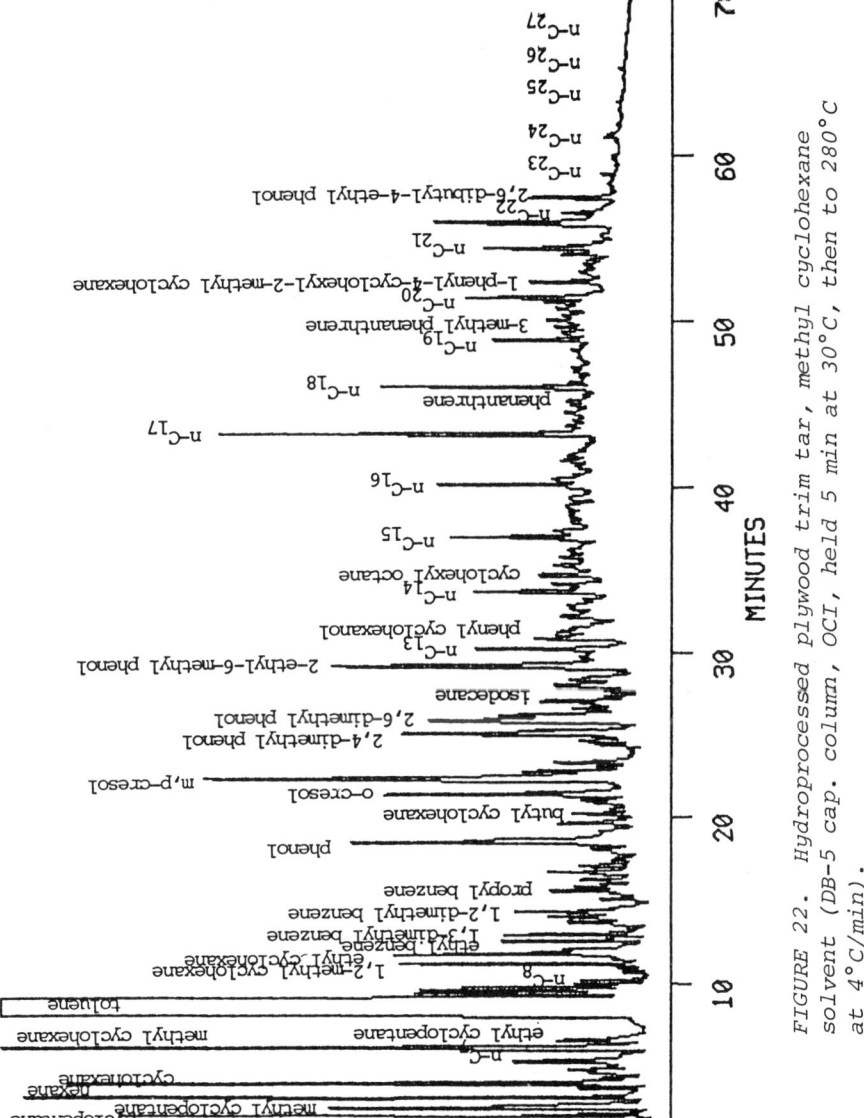

FIGURE 22. Hydroprocessed plywood trim tar, methyl cyclohexane solvent (DB-5 cap. column, OCI, held 5 min at 30°C, then to 280°C at 4°C/min).

For all, similarities in composition suggest that thermochemical conversion of biomass with subsequent hydroprocessing of the tars produced may be a somewhat universal system for producing chemically similar products from physically dissimilar biomass feedstocks. All hydroprocessed tars contain straight-chain hydrocarbons suitable for gasoline and diesel usage, alkyl aromatics which can be used as octane boosters, as well as phenolics which are of interest in wood adhesive applications (Soltes 1983d; Soltes and Lin, 1983b).

GPC of the hydroprocessed Tech-Air tar (decalin solvent) results in the trace given in Figure 23. Compare this trace with that in Figure 11 for the raw tar. Fraction 1 at lower retention volumes is the polymeric content; fraction 2, larger molecules (C_{14} to C_{44} size if hydrocarbons); fraction 3, phenolics; fraction 4, aromatics; and, fraction 5, mostly decalin and tetralin solvents but containing as well some smaller molecules. The composition of the aromatic fraction (No. 4) of the hydroprocessed tar is given in Figure 24. Compare Figure 24 with Figure 12 (aromatics in raw tar as separated by GPC).

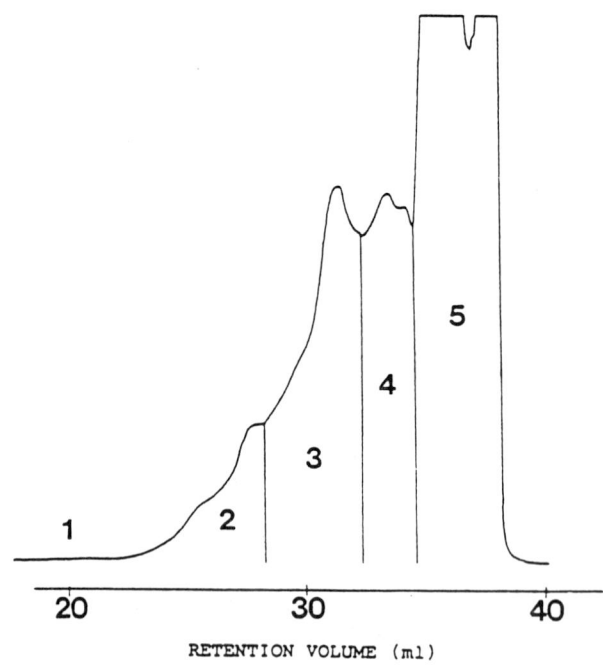

FIGURE 23. GPC of hydroprocessed Tech-Air pyrolytic tar, decaline solvent (Four 30 cm. 10nm micro-styragel columns).

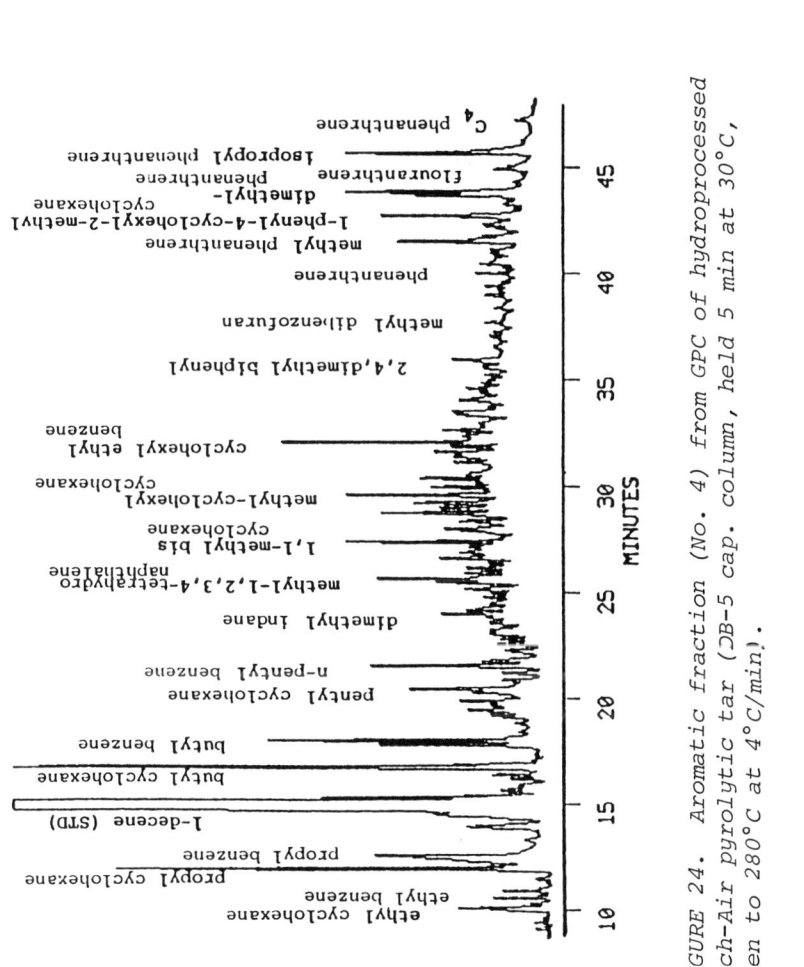

FIGURE 24. Aromatic fraction (No. 4) from GPC of hydroprocessed Tech-Air pyrolytic tar (DB-5 cap. column, held 5 min at 30°C, then to 280°C at 4°C/min).

C. *Fuel Comparisons*

Simulated distillation is a method for determining the boiling point distribution of hydrocarbons using a low resolution gas chromatographic technique. Commercial diesel fuel was used to standardize the method. The hydroprocessed Tech-Air tar product subjected to simulated distillation gave the results shown in Figure 25. This reveals that roughly 40% of the hydroprocessed tar is distributed in the gasoline b.p. range, some 50% in the diesel range with some 10% heavier material.

The hydroprocessed tars themselves, without any fractionation, already exhibit some desirable fuel properties (see Table XII). Upon fractionation, compositions very similar to those of commercial gasoline and commercial diesel oil can be produced (see Figures 26 to 29).

D. *Engine Evaluations of Experimental Fuels*

A meaningful test procedure for experimental fuels is their direct evaluation in small engines equipped with basic dynamometer equipment. An engine can be a valuable experimental tool that can specify to what concentration a conventional fuel can be diluted with experimental fuel before either power losses are observed or operational problems are experienced. Whole hydroprocessed tar can be admixed with diesel fuel in concentrations of up to 25% and used in a small diesel engine without apparent difficulty (Soltes, 1982). At 50 to 75% substitution, increases in fuel consumption and some smoke were observed.

Fractionation of the hydroprocessed tar results in improved engine performance. See Tables XIII and XIV for results of small engine testing with these fractions.

The results shown in Table XIV were encouraging enough that some limited experience has now been had with the use of the light gasoline-like fraction in a 250cc motorcycle. As judged subjectively, the experimental fuel performs well in the motorcycle, with good throttle response and no hesitation in cold start.

After this favorable motorcycle experience, the fuel was subjected to a more rigorous examination in a variable speed test conducted by the Mechanical Engineering Department at Texas A & M. A single cyclinder Christie variable compression engine was used with compression ratio set at 8.8:1, and timing at 21.5° before TDC. Fuel used for this evaluation was a 20% mixture of the light fraction of hydroprocessed tar (< 130°C) in indolene (a research fuel with research octane of 97.3). Results (Figure 30) indicate that the brake

Hydroprocessing of Biomass Tars for Liquid Engine Fuels   51

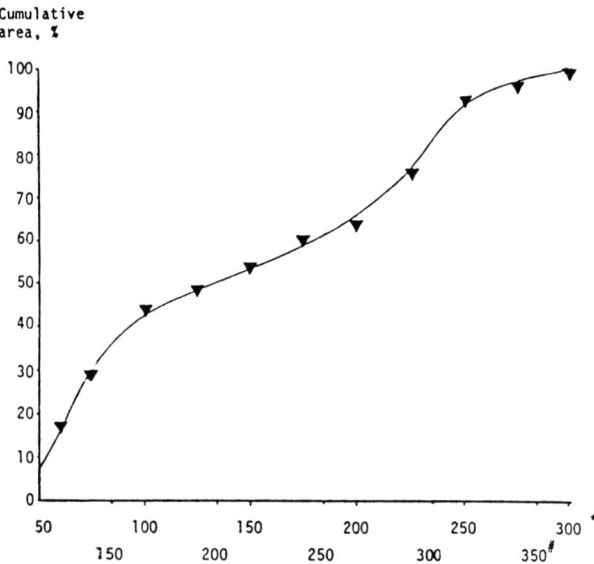

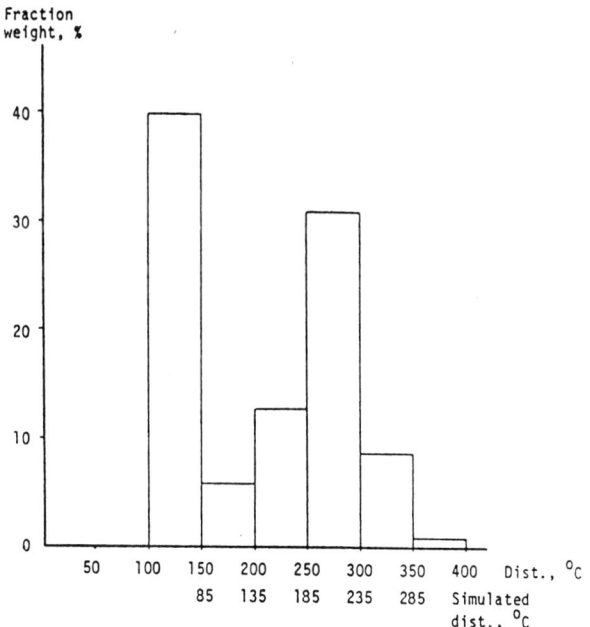

FIGURE 25. (a) Simulated distillation curve for hydro-processed Tech-Air pyrolytic tar; (b) Histogram, fraction distribution. (*) Simulated distillation temperatures; (#) True distillation temperatures.

TABLE XII. Some Properties of Hydroprocessed Tar in Comparison with Those of Commercial Diesel Oil[a]

	Hydroprocessed tar	Diesel oil
Kinetic viscosity, cSt		
at 10°C	2.459	8.062
20°C	2.077	4.253
30°C	1.758	3.400
40°C	1.511	2.877
Heating value MJ/kg	44.67	46.38
Specific gravity	0.824	0.840
Carbon residue, %	1.35	0.34

[a] Hydroprocessed tar here is the Tech-Air product using methyl cyclohexane hydrogen-donor solvent, then stripped to remove solvent.

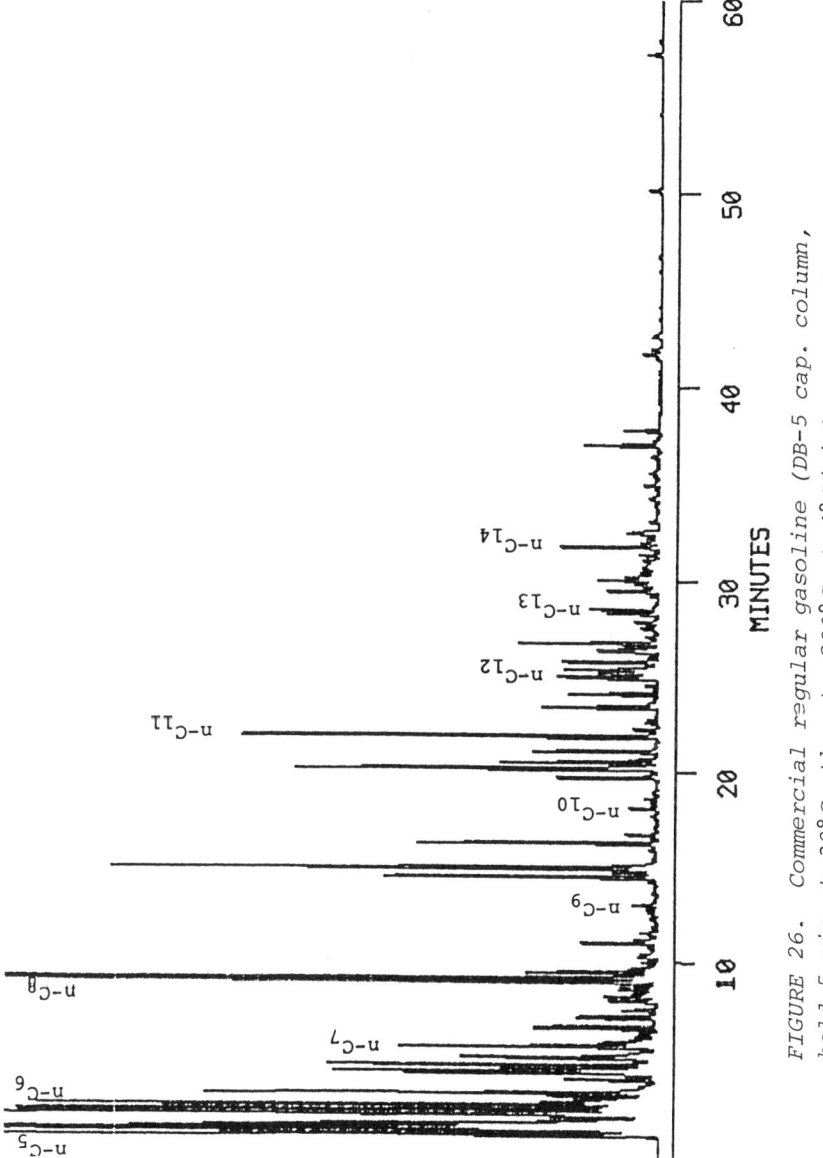

FIGURE 26. Commercial regular gasoline (DB-5 cap. column, held 5 min at 30°C, then to 280°C at 4°C/min).

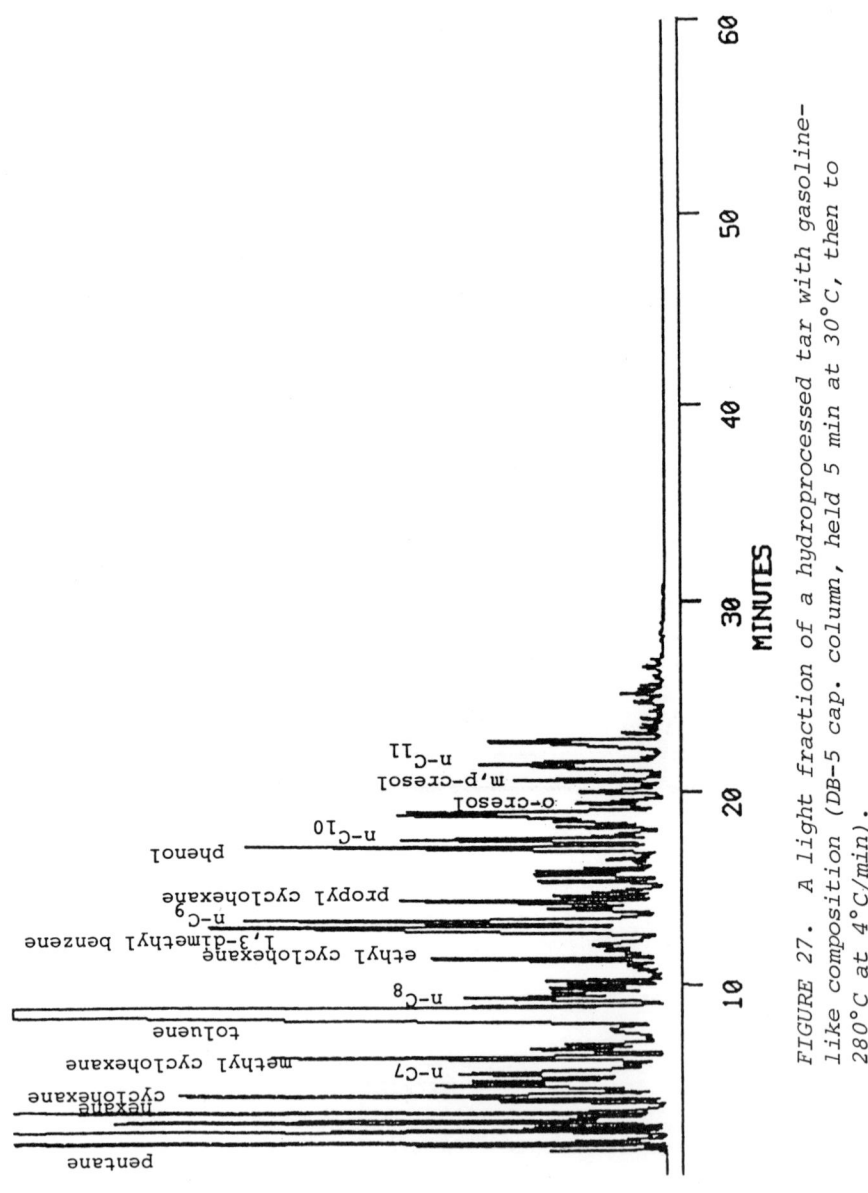

FIGURE 27. A light fraction of a hydroprocessed tar with gasoline-like composition (DB-5 cap. column, held 5 min at 30°C, then to 280°C at 4°C/min).

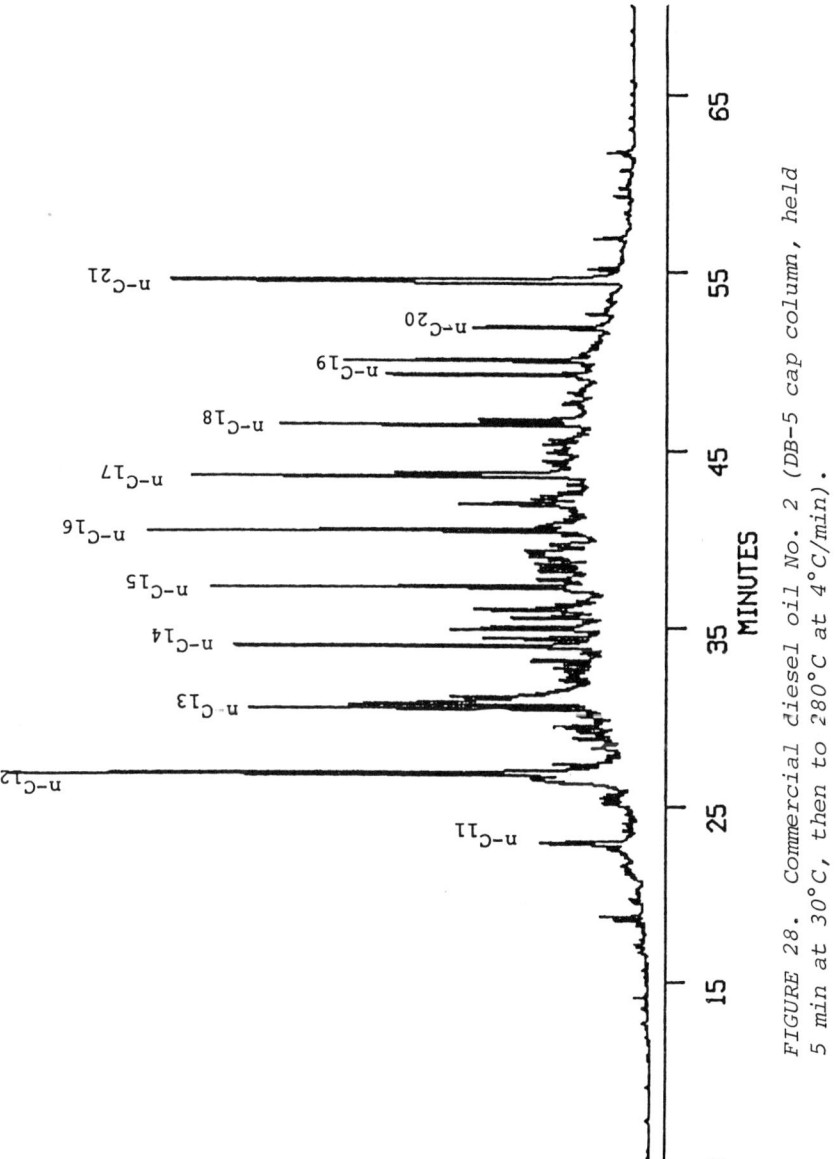

FIGURE 28. Commercial diesel oil No. 2 (DB-5 cap column, held 5 min at 30°C, then to 280°C at 4°C/min).

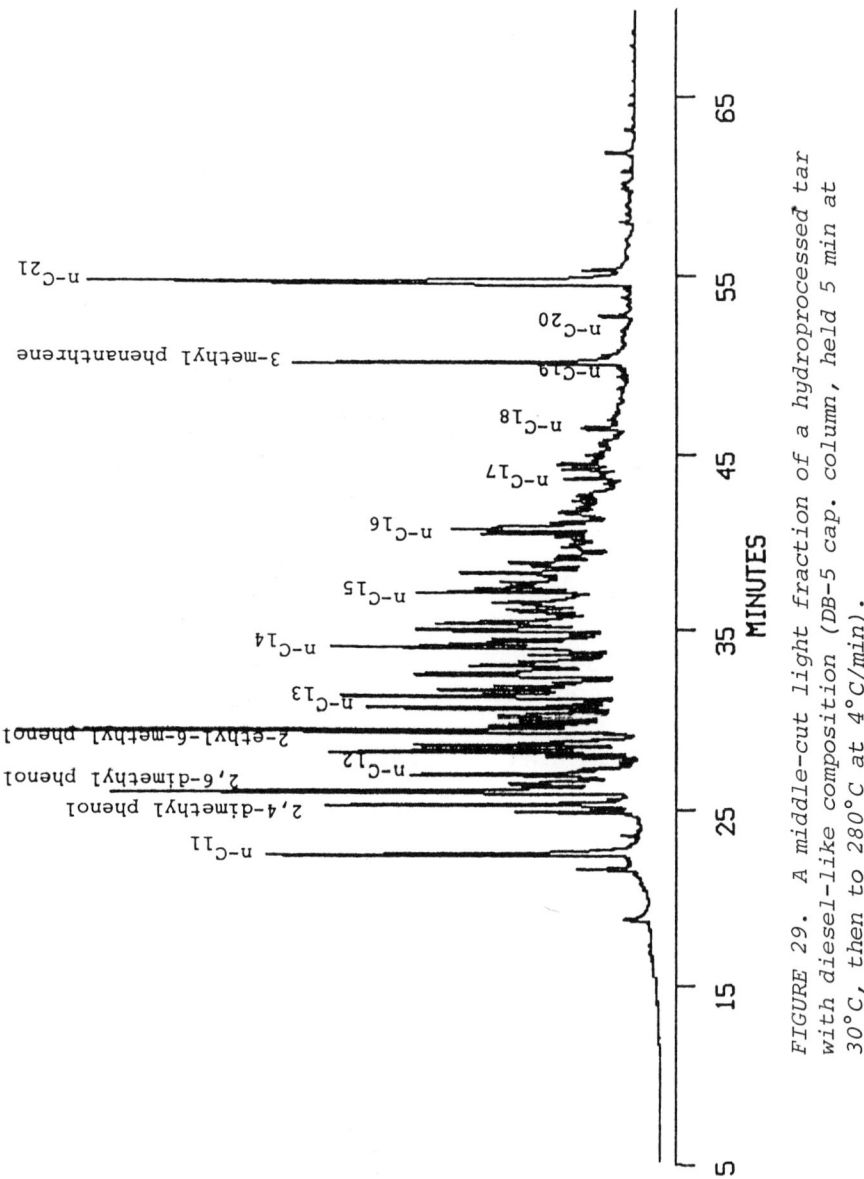

FIGURE 29. A middle-cut light fraction of a hydroprocessed tar with diesel-like composition (DB-5 cap. column, held 5 min at 30°C, then to 280°C at 4°C/min).

TABLE XIII. Small Diesel Engine Performance of Experimental Fuels

Fuel used	Diesel	Diesel + 130-120°C fraction of hydroprocessed tar				
Conventional fuel, %	100	90	75	65	50
Torque, ft. lbs.	1.5	1.5	1.5	1.5	1.5
RPM	2500	2500	2500	2500	2500
Fuel consumption, g.	—	136.4	119.4	111.9	252.0
Time of run, min.	—	20	15	15	26
Heat content of fuel, Btu/g.	42.7	41.9	40.6	39.9	38.6
Efficiency, %	10.2[a]	10.6	9.7	9.6	8.1

[a] Average of several runs. 3.5 hp Petter diesel, outfitted with Go-Power dynamometer. Mechanical efficiency is the ratio of the output power to the rate at which chemical energy of the fuel is consumed by the engine.

TABLE XIV. Small Gasoline Engine Performance of Experimental Fuel[a]

Fuel used	Gasoline		Light < 130°C fraction of hydroprocessed tar			
Conventional fuel, %	100	100	0	0	0	0
Torque, ft. lbs.	1.0	1.0	1.0	1.0	1.0	1.0
RPM	2800	2850	2800	2900	2450	
Fuel consumption, g.	209	209	454	454	215	
Time of run, min.	30	30	60	60	29	
Heat content of fuel, Btu/g.	39.0	39.0	38.6	38.6	38.6	
Efficiency, %	8.3	8.5	7.7	8.0	6.9	

[a] 3.0 hp Briggs & Stratton gasoline engine, outfitted with Go-Power dynamometer.

Hydroprocessing of Biomass Tars for Liquid Engine Fuels

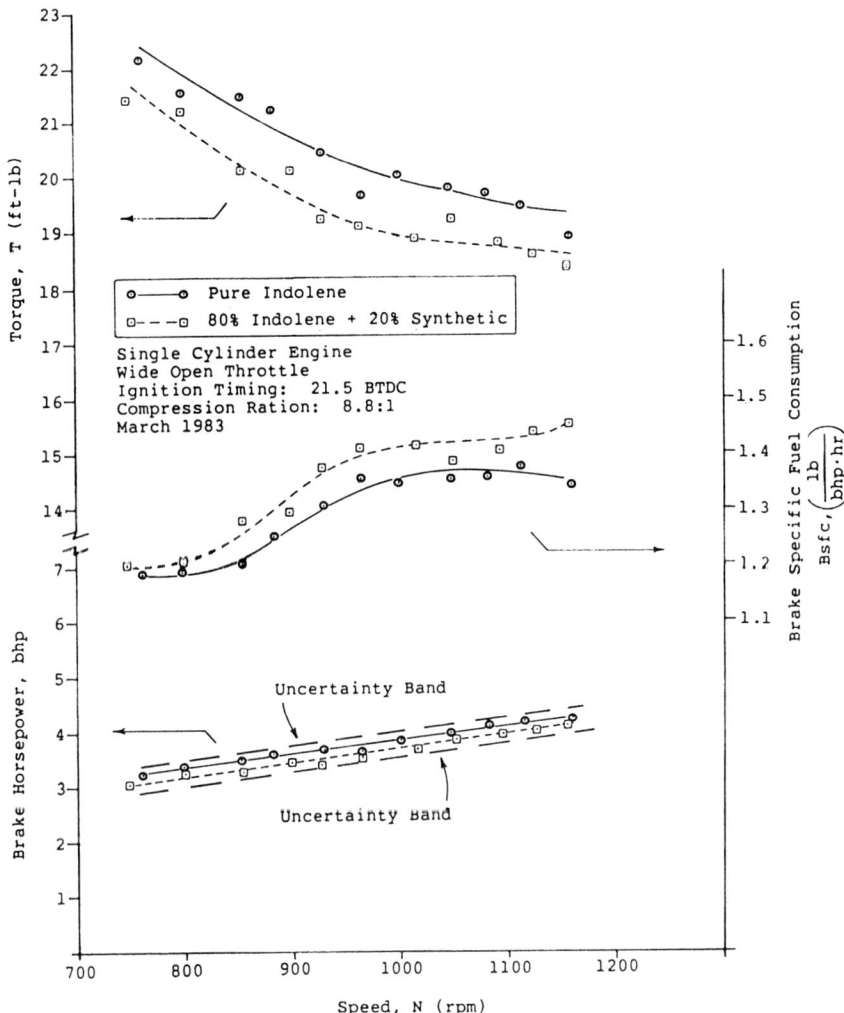

FIGURE 30. Brake horsepower and torque as a function of engine speed for pure indolene and a mixture of 20% by weight light fraction of hydroprocessed tar and 80% indolene.

horsepower produced by both fuels were within experimental error. Brake specific fuel consumption (bsfc) was slightly higher for the mixture, and is accounted for by the higher oxygen content (lower calorific value) of the experimental fuel. By comparison, it may be noted that the use of gasohol (at 10% ethanol) vs. gasoline will result in lower brake horsepower and significantly higher bsfc.

V. PROCESS CONSIDERATIONS

This approach to biomass conversion can serve integrated fuels and chemicals production interests in an overall process concept that makes use of the inherent oxychemical nature of biomaterials with provision for all material and energy requirements from parent biomass feedstocks.

A. Idealized Process Flowsheet

An idealized process flowsheet for the conversion of biomass residues to gasoline and diesel hydrocarbon fuels has been proposed (see Figure 31; Soltes, 1983d). Biomass residues are first pyrolyzed to yield gas, tar and char products (Figure 31a). The gas product, following Tech-Air practice, would probably be combusted to provide heat to dry incoming feed. The tar would be catalytically hydroprocessed (Figure 31b) to yield an oil product which would serve, when fractionated, to yield light ends (recycled for solvent purposes), a middle-cut fuel feedstock, and heavy resids (recycled for recracking). Light ends would be recycled for solvent purposes, and aid as well in washing catalyst in catalyst recovery. Since there is typically no sulfur or other catalyst poison in biomass feedstocks, catalysts can enjoy long service life, an important consideration in the use of noble metal catalysts. A small heavier residual fraction can also be recycled for recracking as shown. The middle-cut fuel feedstock would be further fractionated into raw gasoline and diesel which would be blended with appropriate additives to yield gasoline and diesel fuels (Figure 31c). The gasoline fraction contains the alkyl aromatics, such as toluene and ethyl benzene which are used by the petroleum industry as octane boosters.

Further, the hydrogen content of the chars produced in biomass pyrolysis is sufficient to satisfy the hydrogen requirements of catalytic hydroprocessing, and upon further study, may be available for such use through appropriate

Hydroprocessing of Biomass Tars for Liquid Engine Fuels

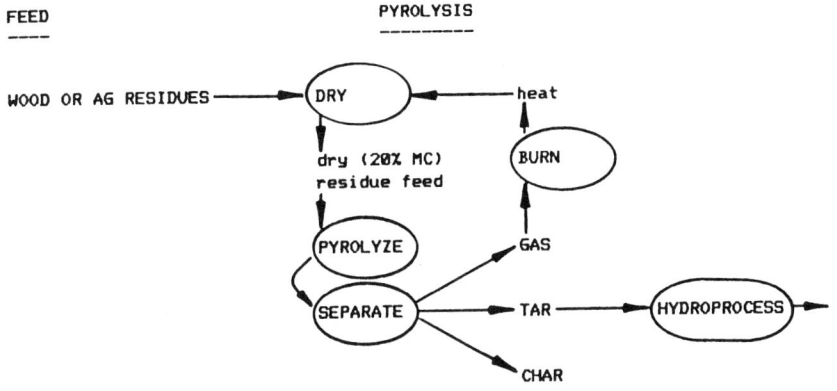

Figure 31-a

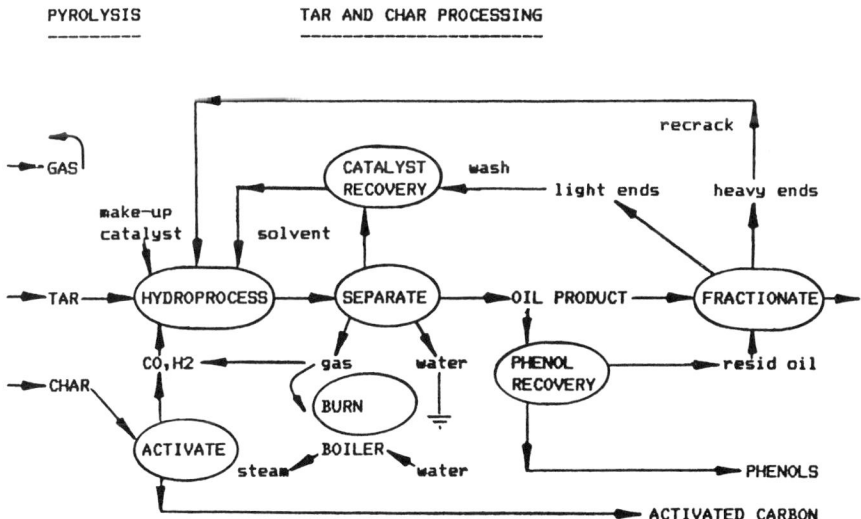

Figure 31-b

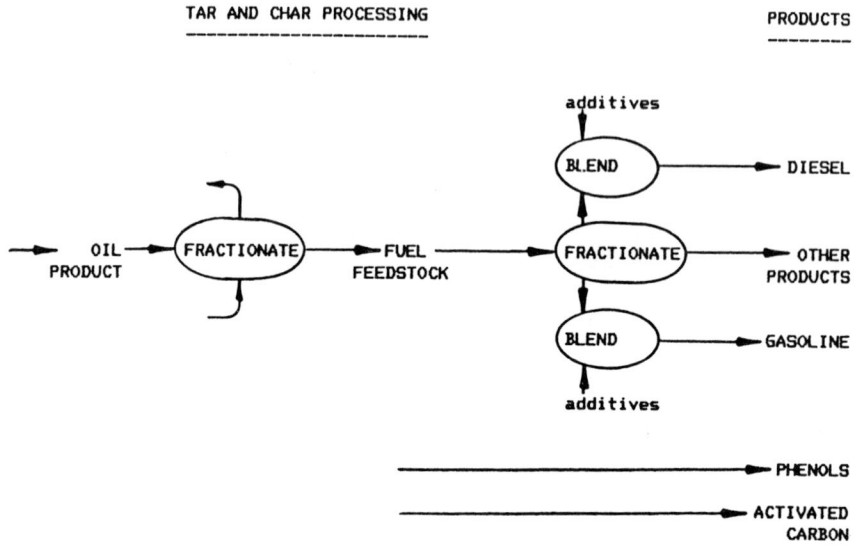

FIGURE 31. Idealized process flowsheet for the production of motor fuels, phenolics and activated carbon from biomass residues.

steam-char reactions. Activated carbons could be generated as byproducts (Figure 31c). Energy requirements for both pyrolysis and subsequent processing may be derived from the gaseous products of pyrolysis. Thus, save for catalyst requirements, all raw material and energy needs to produce liquid hydrocarbon fuels, phenolics and activated carbons can be supplied by a biomass feedstock. To complete the process, an optional phenol recovery step can be entertained. Some catalysts and reaction conditions give mixed hydrocarbons and phenolic products. Beside the use of such phenolics for adhesives (Elder, 1979; Elder and Soltes, 1979b; Soltes, 1980b; Soltes and Lin, 1983b), several phenolics have been promoted as octane boosters (Elliott, 1981), and may additionally serve fuel interests.

B. *Other Process Considerations*

Although it appears that noble metal catalysts give the best results to date in batch processing, work with other types of catalysts has been limited. Further research is required to evaluate other catalysts. Emphasis in catalyst selection should be given to readily available, less expensive catalysts which not only promise general activity in tar hydroprocessing but some specificity towards moderation of hydrotreating activity for coproduct generation.

Catalysts generally available are those that have been developed by the petrochemical industry and used primarily for the conversion of hydrocarbons. The area of catalytic reaction engineering of oxygenated hydrocarbons, such as found in the composition of biomass thermochemical tars, has been largely ignored. More work is required in identifying new catalysts that are nonmetallic based, and which exhibit activities with oxygenated species. There has been much interest of late in the shape-selective zeolites, especially of the Mobil ZSM-5 type (Rabo, 1976; Chen and Garwood, 1978; Weisz et al., 1979) which has specific activity to convert oxygenated species such as methanol and ethanol into hydrocarbon mixtures suitable for motor fuel application. Clays appear to exhibit much potential for catalytic oxychemical conversion, and these in general are being evaluated for their potentials and specificities with biomass tars (Lin and Soltes, 1984).

Further, although the research reported has indicated that gaoline and diesel fuel mixtures may be fractionated out of the whole hydroprocessed tar product, the composition of the latter does not lend itself to high yields of either. Once hydrocarbons are formed, however, conventional reforming catalysts may be employed to specifically convert hydrocarbon mixtures into motor fuel compositions. In current research, reforming catalysts are employed both on tars hydroprocessed with noble metal catalysts and on tars hydroprocessed with novel nonmetallic catalysts. The expectations are that a two-stage hydroprocessing/reforming operation will yield gasoline or diesel fuels more efficiently.

Commercial gasoline and diesel fuels undergo several refining and reforming operations to effect desired compositions and ignition qualities. Gasoline has a boiling point range from approximately 40°C to 160°C, and is a blend of paraffins, naphthalenes and aromatics. There are additional differences between regular gaoline in which octane is improved with tetraethyl lead, and unleaded gasoline, in which aromatics or ethanl (premium unleaded) are used for octane enhancement. There are additionally additives for a host of important performance properties, such as oxidation inhibitors, and addition of lower hydrocarbons in winter months for improved cold starts, etc. Other reformed hydrocarbons are also blended into gasoline for performance properties. For gasoline, aromatics are better fuels than the paraffins. Diesel, however, is a mixture of paraffins, differing in many compositional features depending on the petroleum feedstock source, and for which cetane index is controlled between 35 and 50 with composition and additives, such as amyl nitrate. For diesel fuels, paraffins give better ignition properties than aromatics. Since the hydroprocessed tars contain naphthalenes and aromatics, it is understandable why the gasoline fractions work better than the diesel fractions. Improvements in diesel performance must be addressed through post-processing operations which effect ring opening reactions.

C. *Process Definition*

The overall goal of this program is to develop a process which can convert agricultural and forestry residues into higher valued energy and chemical commodity products. It is perceived that biomass conversion processes are needed in which some of the chemical complexity of the original biomass composition is first retained in product values not readily obtained from other sources (if not carbohydrate or phenolic polymers, then at least monomeric oxy-chemicals), with only the balance unusable for this purpose relegated to fuel status. The two-stage process of biomass pyrolysis followed by hydroprocessing of the tar produced, with needed hydrogen supplied by steam/char reactions, has potential for converting the residues of biomass conversion processes into such products: oxy-chemicals (especially reactive phenolics if hydrotreating activity is moderated) and activated carbons, with the balance processed into liquid internal combustion fuels. When the primary tar hydroprocessing process has been modeled (Sheu and Soltes, 1984), and further information is obtained on the composition of products hydroprocessed under differing reaction conditions (Lin and Soltes, 1984), it is expected that various alternatives in product mix will have been identified which will form the basis for continuing process development studies. Particular attention will be paid to hydrotreating activity in hydroprocessing towards the coproduction of oxy-chemical products for their potential incorporation into some definitive process.

VI. CONCLUSIONS

This research program was directed towards establishing to what extent the thermochemical conversion of biomass with subjequent hydroprocessing of the tars produced may produce functionally similar fuel and chemical products from physically dissimilar biomass feedstocks. It was determined that the desired process universality does exist.

It is reported that both the tars produced from a variety of agricultural residues and their hydroprocessed products are indeed similar in compositions. It appears that although the physical forms of such residues are different, tars produced via thermochemical processes exhibit many similarities in chemical composition, and that many differences in the compositions of different tars are virtually eradicated and catalytic hydroprocessing. Similar products were produced from Tech-Air pine pyrolysis tar, DeKalb corncob gasification tar, and the tars produced in a gasification/pyrolysis reactor from wood chips, pecan shell, sugarcane bagasse, peanut shell, corncob, rice hull, cottonseed hull, and plywood trim. Fuels derived from the tars are physically and chemically similar to gasoline and diesel fuels. A process flowsheet is described which, except for catalyst, appears to be energy and materials self-sufficient in the production of liquid hydrocarbons, phenolics and activated carbons.

Although there has been much process research and demonstration activity in biomass pyrolysis, and although pyrolysis has been promoted as a biomass conversion process, there has been no implementation on a commercial scale. The major problem may be effective utilization of all three primary products. The processes described should impact favorably on the economical utilization of biomass as an alternative energy source for the production of motor fuels and oxychemicals.

It is ventured that the application of hydroprocessing conditions to any tars from biomass will result in similar hydrocarbon compositions. If this can be confirmed in future investigations, then a renewable route from biomass to aromatic and paraffinic petroleum feedstocks can be claimed.

REFERENCES

Antal, M. J. 1983. Biomass Pyrolysis: A Review of the Literature. Part 1 - Carbohydrate Pyrolysis. In "Advances in Solar Energy - 1982." American Solar Energy Society, New York.

Biggs, W. A., Wise, J. T., Cook, W. R., Baxley, W. H., Robertson, J. D., and Copenhaver, J. E. 1961. Tappi 44: 386.

Bosich, E. 1970. "Corrosion Prevention for Practicing Engineers." Barnes and Noble, New York.

Chatterjee, A. K. 1981. State-of-the-Art Review on Pyrolysis of Wood and Agricultural Biomass. Final report, Contract No. 53-319-R-0-206, AC Project P0380, USDA Forest Service, Washington, D.C.

Chen, N. Y., and Garwood, W. E. 1978. *J. Catal.* 52:453.

Davis, H. G., Kloden, D. J., and Schaleger, L. L. 1981 *Biotechnol. Bioeng. Symp.* 11:151.

Diebold, J. P., and Scahill, J. W. 1983. Entrained Flow, Fast Ablative Pyrolysis of Biomass. SERI/PR-234-2144. Solar Energy Research Institute, Golden, CO.

Elder, T. J. 1979. The Characterization and Potential Utilization of the Phenolic Compounds Found in a Pyrolytic Oil. Ph.D. Dissertation, Texas A&M University.

Elder, T. J., and Soltes, E. J. 1979a. Further Investigations into the Composition and Utility of a Commercial Wood Pyrolysis Oil. Paper presented at the American Chemical Society National Meeting, Honolulu, HI.

Elder, T. J., and Soltes, E. J. 1979b. Adhesive Potentials of Some Phenolic Constituents of Pine Pyrolytic Oil. Paper presented at the American Chemical Society National Meeting, Washington, D.C.

Elder, T. J., and Soltes, E. J. 1980. *Wood & Fiber* 12:217.

Elliott, D. C. 1981. *Biotechnol. Bioeng. Symp.* 11:187.

Knight, J. A., Bowen, M. D., and Purdy, K. R. 1976. Pyrolysis - A Method for Conversion of Forestry Wastes to Useful Fuels. Paper presented at the Forest Products Research Society Meeting, Atlanta, GA.

Lin, S.-C. K. 1978. Volatile Constituents in a Wood Pyrolysis Oil. M.S. Thesis, Texas A&M University.

Lin, S.-C. K. 1981. Hydrocarbons via Catalytic Hydrogen Treatment of Pine Pyrolytic Oil. Ph.D. Dissertation, Texas A&M University.

Lin, S.-C. K., and Soltes, E. J. 1984. Unpublished results.

Rabo, J. A. 1976. "Zeolite Chemistry and Catalysis." American Chemical Society Monograph 171. ACS, Washington, D.C.

Robinson, J. S. 1980. "Fuels from Biomass: Technology and Feasibility." Noyes Data Corp., Park Ridge, NJ.

Sheu, Y.-H. E., Philip, C. V., Anthony, R. G., and Soltes, E. J. 1984. Separation of Functionalities in Pyrolytic Tar by GPC. Submitted to *J. Chromatog*.

Sheu, Y.-H. E., and Soltes, E. J. 1984. Unpublished results.

Soltes, E. J. 1980a. Thermal Conversion of Lignocellulosics: Retrospect and Prospect. Paper presented at the Intersciencia/JSST Seminar and Workshop on "Materials for the Future: Renewable Organic Resources for Industrial Materials," Kingston, Jamaica., Nov.

Soltes, E. J. 1980b. *Tappi* 63:75.

Soltes, E. J. 1982. Diesel Fuels from Pine Pyrolytic Oil. Final Report, USDA Energy Grants FY 1980, Project 59-2481-0-2-189-0.

Soltes, E. J. 1983a. Thermochemical Routes to Chemicals, Fuels and Energy from Forestry and Agricultural Residues. *In* "Biomass Utilization" (W. A. Cote, Jr., ed.). Plenum Press, New York.

Soltes, E. J. 1983b. Biomass Thermal Degradation Tars as Sources of Chemicals and Fuel Hydrocarbons. *In* "Wood and Agricultural Residues: Research on Use for Feed, Fuels and Chemicals" (E. J. Soltes, ed.). Academic Press, New York.

Soltes, E. J. 1983c. Liquid Engine Fuels from Biomass Pyrolytic Tars. Final Report, USDA Energy Grants FY 1980, Project 59-2481-0-2-189-0.

Soltes, E. J. 1983d. *Appl. Polymer. Symp.* 37:775.

Soltes, E. J., and Elder, T. J. 1981. Pyrolysis. *In* "Organic Chemicals from Biomass" (I. S. Goldstein, ed.). CRC Press Inc., Boca Raton, FL.

Soltes, E. J., and Lin, S.-C. K. 1983a. Vehicular Fuels and Oxychemicals from Biomass Pyrolytic Tars. *Biotechnol. Bioeng. Symp. 13*. In press.

Soltes, E. J., and Lin, S.-C. K. 1983b. Adhesives from Natural Resources. *In* "Progress in Biomass Conversion, Volume 4" (D. A. Tillman and E. C. Jahn, eds.). Academic Press, New York.

Soltes, E. J., Wiley, A. T., and Lin, S.-C. K. 1981. *Bioeng. Symp.* 11:125.

Tatom, J. W., Wellborn, H. W., Harahap, F., and Sasmojo, S. 1980. Third World Applications of Pyrolysis of Agricultural and Forestry Wastes. *In* "Thermal Conversion of Solid Wastes and Biomass" (J. L. Jones and S. B. Radding, eds.). American Chemical Society Symposium Series *130*. ACS, Washington, D.C.

Weisz, P. B., Haag, W. O., and Rodewald, P. G. 1979. *Science 206*:57.

ACKNOWLEDGMENTS

The major portion of this paper is based on work supported by the U.S. Department of Agriculture under Special Research Grants Program Agreement Nos. 59-2481-0-2-089-0 and 59-2481-1-2-123-0. Other work was sponsored by the Center for Energy and Mineral Resources, and the Texas Agricultural Experiment Station, both parts of the Texas A & M University System. Any opinions, findings, conclusions or recommendations expressed in this article are those of the authors and do not necessarily reflect the views of the sponsoring agencies. Special acknowledgement is made to the following people involved in various phases of this research program: K. Allmand, T. Elder, L. Hughes, S. Kuh, J. Lanahan, P. Ouyang, P. Parris, D. Shah, E. Sheu, C. Stauder, R. Summers, A. Wiley, and J. Wolfhagen.

FUEL CHARACTERISTICS OF WOOD AND NONWOOD BIOMASS FUELS

Amadeo Rossi

Envirosphere Company
Bellevue, Washington

I. INTRODUCTION 70

II. METHODOLOGY 71

    A. Proximate Analysis 72

    B. Ultimate Analysis 73

    C. Higher Heating Values 73

    D. Bulk Density 74

    E. Moisture Content 74

    F. Screen Fractionation 75

    G. Ash Fusion Temperature 75

III. RESULTS 75

    A. Proximate Analysis 75

    B. Ultimate Analysis 78

    C. Higher Heating Values 79

    D. Bulk Density 79

            E. Screen Fractionation   79

            F. Ash Fusion Temperature   79

       IV. DISCUSSION 87

            A. Chemical Characterization of Biomass Fuels .   87

            B. Existing Combustion Facilities   95

        V. CONCLUSIONS 96

I. INTRODUCTION

    Traditionally, wood residues generated from the forest
products industry were disposed of either as a solid waste or
were burned without energy recovery to reduce their volume
before land disposal. The traditional mark of this type of
combustion process was the teepee burner. Today teepee burners are almost nonexistent. Today pulp mills, sawmills, and
plywood mills, as well as large integrated mill complexes and
stand alone wood fired power plants, are reaching out hundreds
of miles for both raw material and fuel, virtually consuming
all available mill residues. As a result, mill residues
represent a marketable product with a premium value. Because
of this drain on mill residues, wood fuel has become valuable
enough for contractors to enter the woods for the purpose of
producing fuel.
    Another source of biomass residues is the agricultural
industry. Currently, in the agricultural industry the
residues are not being used to the extent they are in the forest products industry. However, the potential exists. Further, these materials could be used by entrepreneurs for the
generation of electricity in stand-alone biomass fired power
plants rather than, or as a supplement to, wood. As with the
forest products industry, traditionally agricultural and processing residues were either field burned or landfilled. However, today it is being realized that these residue materials
could significantly affect the economic viability of agricultural firms as well as food processing facilities. Examples
of processing residues include peach and olive pits from de-
pithing operations; rice hulls from the processing of rice;
trash from the cotton ginning operation; tomato pomace from

the processing of tomatoes into canned products and catsup; and grape pomace from winemaking. Agricultural residues include vineyard prunings, grasses, and stone fruit prunings. The word "wood" will be used here to include only traditional forest products industry mill residues and forest residuals, not agricultural derived residues.

In the forest products industry, energy derived from wood is used to generate electricity for process drives and infrared drying operations and steam for dry kilns, veneer dryers, paper machines, digestors, evaporators, log vats, hot presses, plant heat, and paint lines. This energy demand has traditionally been met by large low pressure boilers without superheat. Today many such units are being replaced by modern cogeneration and condensing facilities.

Similar needs exist in the agricultural community as well as at food processing facilities, where steam is still the major source of heat and electricity is used for process drives. Some uses for process steam in the food processing industry include coolers/chillers; evaporators; concentrators; atmospheric and pressurized cookers; production of hot water for process, cleanup, and sanitation needs; plant heat; and replacement of infrared or natural gas lacquer lines in can manufacturing operations.

Before agricultural residues can be combusted to produce such energy, the chemical and physical properties of these fuels must be evaluated. The fuel resource is the single most important factor in the design of biomass fired combustors. The purpose of this chapter, therefore, is to discuss the chemical and physical properties of agricultural and wood residues with respect to direct combustion and gasification systems. To evaluate the chemical and physical properties of these fuels, specific fuel tests are employed. These tests include proximate analysis, ultimate analysis, higher heating value, bulk density, moisture content, screen fractionation, and ash fusion temperature. The values presented for these tests are representative, with no assigned statistical significance to the universe of samples. The presentation of the test results will focus on the chemical and physical properties of the material as applied to combustor design and environmental concerns.

II. METHODOLOGY

The values obtained from the fuel tests will be used as the basis for the comparison of the combustion properties of wood and agricultural type fuels. The following is a brief overview of each test to determine the combustion properties of various biomass fuels.

A. Proximate Analysis

Proximate analysis determines the percentage of volatile matter, fixed carbon, and ash in a particular biomass fuel. These results give an indication of the degree of flaming and glowing combustion. The proximate analyses conducted are based on ASTM Standard D3172-73, titled "Proximate Analysis of Coal and Coke." This standard was followed on a procedural basis.[1] However, the temperatures and times are adjusted to take into account the higher volatile to fixed carbon ratio of biomass fuels. The different temperatures and times used are given below (Mingle and Boubel, 1968).

(1) The coal furnace temperature used is 1760 ± 210°F (960 ± 20°C);

(2) A flaming time of 1 minute;

(3) A soaking time of 6 minutes;

(4) A muffle furnace temperature of 1380 ± 210°F (750 ± 20°C);

(5) An ashing time of 1 1/2 hours; and

(6) A moisture content of zero.

The difference in weight of the material before and after the flaming and soak time (i.e., 7 minutes) is defined as the amount of volatile matter while the material left after 1 1/2 hours of ashing is the amount of ash. The difference in weight of the residue before and after ashing is defined as the amount of fixed carbon.

Equipment used for the proximate analysis includes a Hoskins crucible furnace (Model FA-120) with an autotransformer, a calibrated digital readout thermocouple (Model W5M), and a Thermolyne muffle furnace with automatic temperature control (Model F-D1525M).

---

[1]*Except the fixed carbon sample was ashed rather than a separate sample to improve the level of accuracy of the results.*

B. Ultimate Analysis

The ultimate analysis is the experimental determination of the amount of hydrogen, carbon, oxygen, sulfur, nitrogen, and ash present in a particular fuel sample. Based on the results of the ultimate analysis, the hydrogen-to-carbon and oxygen-to-carbon ratios can be calculated, giving an indication of the reactivity of a particular fuel. An empirical formula and the combustion equation for a particular fuel can also be calculated.

All of the ultimate analyses are conducted by Hazen Research in Colorado, using ASTM Standard D3180-79 titled "Ultimate Analysis of Coal and Coke." The samples are dried to oven dry weight and a residual moisture content is then conducted as part of the ultimate analysis. The sample size is 14 grams of 60 mesh material.

C. Higher Heating Values

The higher heating value (HHV) of a particular fuel is a measure of the amount of energy a fuel can release under ideal combustion conditions in a pure oxygen environment. The HHV is an important parameter to obtain in order to calculate the amount of a particular fuel required for a given application. The HHV includes the latent heat of vaporization of the liquid water produced by the combustion of hydrogen. The net or lower heating value (LHV) does not include this latent heat since the water remains in the vapor phase.

The higher heating values are obtained using a fully automatic adiabatic bomb calorimeter manufactured by Parr (Model 1241). The procedure followed is that set forth by ASTM Standard D2015-77 titled "Gross Calorific Value of Solid Fuel by the Adiabatic-Jacket Bomb Calorimeter." Good burns are obtained without pelletizing the fuel unless the specific gravity of the sample is low. Low specific gravity samples are pelletized or a reduced sample size is used. Also, the correction factors for the fuse wire and the formation of sulfuric acid and nitric acid are ignored. The fuse wire correction factor is ignored since 36 gage platinum wire is used rather than the conventional nickel-chromium wire. The correction factors for sulfuric and nitric acid formation are ignored because of the low sulfur and nitrogen content of biomass fuels. An experimental error is calculated for ignoring these correction factors, based on a fuel with a 0.1% and 0.5% sulfur and nigrogen content, respectively. The experimental error calculated is 0.13% of the HHV for sulfur and 0.43% of the HHV for nitrogen, based on the heats of formation and reaction. This error appears within the limits of calibration using standard benzoic acid tablets.

D. *Bulk Density*

The bulk density is another important fuel parameter. The bulk density is the density of the gross fuel sample and has the units of lbs/ft^3 (kg/m^3). The bulk density plays an important role in the design of the combustor fire box as well as the ultimate choice of a combustion method. It is equally important in the design of mechanical and pneumatic conveying systems used to feed the combustor.

The values of the bulk density are a function of the source and method of preparation of a fuel sample, the quantity of sample taken, placement of the sample into the container, and the sample moisture content. As expected, there can be a tremendous variation between samples. For example, Douglas-fir shavings, sanderdust, and sawdust all have significantly different bulk densities even though they have the same specific gravity.

To facilitate comparison of bulk densities between widely different fuels the values presented here are all for samples that pass a 0.0787 inch (2 mm) screen and remain on a 0.0394 inch (1 mm) screen. Thus, all fuels are compared on an equal basis.

E. *Moisture Content*

The moisture content is not determined for each biomass fuel due to the wide range of moisture contents possible. As with bulk density, moisture content is very site specific. However, moisture content is an important parameter that deserves careful evaluation for any combustion process. Further, the moisture content affects the "as-fired" heat content of the fuel as well as the thermal efficiency.

Although moisture contents are not calculated for each fuel, they are postulated for different types of fuel based on experience. All moisture contents are presented on a green basis. That is, they are defined as moisture content green, MCg. The value of MCg in percent is obtained by dividing the water in a specific sample by its green or "as is" weight and multiplying that by 100 to convert it to percent. In order to calculate the amount of water in a fuel sample, it is dried in a convection oven until constant weight, usually 8 to 24 hours depending on the sample (Browning, 1967).

F. Screen Fractionation

Representative screen fractionations and bulk densities are presented to characterize the percent fines and their ash content for selected fuels. However, screen fractionation is not considered in detail, due to the variations encountered in such analysis based on the fuel source and method of preparation. Factors which affect a fuels screen fractionation include the operational characteristics of a chipper, hog, or hammermill; the degree of screening; amount of dirt; moisture content, and source of generation. All of these factors need to be addressed before any meaningful analysis can be conducted.

G. Ash Fusion Temperature

The ash fusion temperature measures the degree of deformation of a cone of ash at four different temperatures as defined in ASTM Standard D1857-68, titled "Fusibility of Coal and Coke Ash." The four different temperatures recorded are the temperature of initial deformation, softening, hemispherical, and the fluid temperature. As a result, the ash fusion temperature is an indication of the potential of slagging expected. Representative ash fusion temperatures are conducted by Hazen Research.

III. RESULTS

A summary of experimental results for the various fuel tests evaluated are presented in this section. From these data certain conclusions can be drawn regarding the various types of biomass fuels.

A. Proximate Analysis

The proximate analysis along with the volatile-to-fixed carbon ratio of agricultural and wood residues are presented in Table I and Table II, respectively. As indicated in Section II, the moisture contents are not presented on a fuel specific basis. A general classification of the moisture contents on a green basis of selected fuels is presented in Table III.

TABLE I. Proximate Analysis of Various Agricultural Fuels Including the Volatile to Fixed Carbon Ratio

Fuel	Volatiles %	Fixed carbon %	Ash %	Volatile to fixed carbon ratio
Cherry pits	84.2	14.8	1.0	5.7
Coconut fiber	69.6	25.3	5.1	2.8
Coconut shell	81.8	17.6	0.6	4.6
Gin trash	75.4	15.4	9.2[a]	4.9
Gin trash fines	68.2	9.9	21.8	6.9
Grape pomace	74.4	21.4	4.2	3.5
Grape vines	80.1	17.7	2.2	4.5
Mote trash	79.3	14.8	5.9	5.4
Olive pits	80.0	16.9	3.1	4.7
Orchard prunings	83.3	14.6	2.1	5.7
Peach pits	79.1	19.8	1.1	4.0
Peanut shells	71.6	22.2	6.2	3.2
Pistachio nuts	82.7	15.0	2.3	5.5
Plum pits	85.3	14.6	0.1	5.8
Rice hulls	63.6	15.8	20.6	4.0
Tomato pomace	85.1	10.8	4.1	7.9
Vineyard fines	83.4	12.9	3.7	6.5
Vineyard prunings	74.9	14.3	10.8[a]	5.2
Walnut shells	81.2	17.4	1.4	4.7

[a] High ash content probably due to embedded dirt associated with collection and storage of the material.

TABLE II. Proximate Analysis of Various Wood Fuels Including the Volatile to Fixed Carbon Ratio

Fuel	Volatiles %	Fixed carbon %	Ash %	Volatile to fixed carbon ratio
Wood				
Black oak	85.6	13.0	1.4	6.6
Big leaf maple	87.9	11.5	0.6	7.6
Canyon live oak	88.2	11.3	0.5	7.8
Chinkapin	86.9	12.8	0.3	6.8
Douglas-fir	87.3	12.6	0.1	6.9
Madrone	84.5	15.1	0.3	5.5
Red alder	87.1	12.5	0.4	7.0
Tan oak	87.1	12.4	0.5	7.1
Western hemlock	87.0	12.7	0.3	6.8
Western red cedar	86.5	13.2	0.3	6.6
Bark				
Black oak	81.0	16.9	2.1	4.8
Chinkapin	80.5	19.3	0.2	4.2
Cottonwood	79.5	17.5	3.0	4.5
Douglas-fir	73.6	25.9	0.5	2.8
Red alder	77.3	19.7	3.0	3.9
Tan oak	76.3	20.8	2.9	3.7
Western hemlock	73.9	24.3	0.8	3.0
Western red cedar	77.6	21.2	1.2	3.7
Cull material				
Douglas-fir	89.5	10.4	0.1	8.6
Douglas-fir[a]	85.7	14.2	0.1	6.0
Madrone	87.8	12.0	0.2	7.3
Red alder[a]	84.3	15.2	0.5	5.5
Tan oak	90.6	9.2	0.2	9.9

[a] Advanced decay.

TABLE III. Range of Moisture Contents of Selected Biomass Fuels

Fuel	Range of moisture contents (%)
Wood	
Bark	30 - 60
Chips	40 - 50
Cull material	40 - 70
Hog fuel	30 - 60
Planer shavings	8 - 19
Sanderdust	2 - 6
Sawdust	40 - 55
Agricultural	
Gin trash	7 - 12
Grape pomace	50 - 60
Nuts	10 - 35
Orchard prunings	20 - 40
Peach pits	30 - 40
Rice hulls	7 - 10
Tomato pomace	50 - 75
Vineyard prunings	20 - 40

B. Ultimate Analysis

The ultimate analysis of selected agricultural and wood residues are presented in Table IV on a dry weight percent basis. Also included are the calculated values of the hydrogen to carbon atomic ratio (H/C) and the oxygen to carbon atomic ratio (O/C).

C. Higher Heating Values

The higher heating values of selected agricultural and wood fuels are presented in Table V and Table VI, respectively. These values can be calculated on an "as-fired" basis by assuming a specific moisture content. For example, Douglas-fir sawdust with a moisture content of 30% would have an as-fired heat content of 9,790 (22.77 MJ/kg) × 0.70 or 6,853 Btu/lb (15.94 MJ/kg) ignoring the formation of water from the hydrogen in the fuel.

D. Bulk Density

The bulk density of selected agricultural and wood fuels is presented in Table VII. As indicated previously, the bulk density is calculated for material that passes a 0.0787 inch (2 mm) screen but remains on a 0.0394 inch (1 mm) screen). Also presented for a few samples in Table VII is the bulk density of selected ash samples.

E. Screen Fractionation

Screen fractionation data for samples of dry orchard prunings, vineyard prunings, and gin trash as taken from existing fuel piles of combustion devices are presented in Table VIII as well as the bulk density of their respective size fractions. These samples were screened using a 0.0787 inch (2 mm) and 0.0197 inch (500 µm) screen.

F. Ash Fusion Temperature

Results of ash fusion temperature determinations for both oxidizing and reducing atmospheres on vineyard prunings, tan oak, and madrone are presented in Table IX. Vineyard prunings were broken into a dirt fraction, the as received chip, and a chip which was scraped clean.

TABLE IV. Ultimate Analysis of Selected Agricultural and Wood Fuels Including the Hydrogen to Carbon and the Oxygen to Carbon Atomic Ratios

Fuel	Carbon (%)	Hydrogen (%)	Oxygen (%)	Nitrogen (%)	Sulfur (%)	Ash (%)	H/C	O/C
Agricultural								
Almond shells	51.25	6.00	40.01	0.23	0.10	2.41	1.40	0.59
Gin trash	42.77	5.08	35.38	1.53	0.55	14.69	1.43	0.62
Grape pomace	54.94	5.83	32.08	2.09	0.21	4.85	1.27	0.44
Orchard prunings	49.15	5.95	43.24	0.25	0.04	1.38	1.45	0.66
Peach pits	49.14	6.34	43.52	0.48	0.02	0.50	1.55	0.66
Rice hulls	38.30	4.36	35.45	0.83	0.06	21.00	1.36	0.69
Vineyard prunings	47.99	5.65	39.61	0.86	0.08	5.81	1.41	0.62

Wood							
Black oak	48.97	6.04	43.48	0.15	0.02	1.34	1.48	0.67
Big leaf maple	49.89	6.09	43.27	0.14	0.03	0.58	1.47	0.65
Canyon live oak	47.84	5.80	45.76	0.07	0.01	0.52	1.46	0.72
Chinkapin	49.68	5.93	44.03	0.07	0.01	0.28	1.43	0.67
Douglas-fir	50.64	6.18	43.00	0.06	0.02	0.10	1.46	0.64
Douglas-fir cull	52.13	6.11	41.56	0.07	0.01	0.12	1.41	0.60
Madrone	48.56	6.05	45.08	0.05	0.02	0.24	1.50	0.70
Madrone cull	48.94	6.03	44.75	0.05	0.02	0.21	1.49	0.73
Red alder	49.55	6.06	43.78	0.13	0.07	0.41	1.47	0.66
Tan oak	48.34	6.12	44.99	0.03	0.03	0.49	1.52	0.70
Tan oak cull	48.67	6.03	44.99	0.06	0.04	0.20	1.49	0.69

TABLE V. *The Higher Heating Values of Selected Agricultural Fuels*

Fuel	Higher heating value Btu/lb (MJ/kg)
Cherry pits	9,350 (21.75)
Coconut pith and fiber	8,270 (19.24)
Coconut shell	8,840 (20.56)
Gin trash	6,700 (15.58)
Grape pomace	9,373 (21.81)
Grape vines	8,920 (20.95)
Mote trash	7,190 (16.72)
Olive pits	8,330 (19.37)
Orchard prunings	8,190 (19.05)
Peach pits	8,350 (19.42)
Peanut shells	7,540 (17.54)
Pistachio nuts	7,910 (18.40)
Plum pits	9,090 (21.14)
Rice hulls	6,400 (14.89)
Tomato pomace	10,220 (23.77)
Vineyard prunings	7,220 (16.79)
Walnut shells	8,390 (19.51)

TABLE VI. The Higher Heating Values of Selected Wood Fuels

Fuel	Higher heating value Btu/lb (MJ/kg)	
Wood		
Black oak	8,020	(18.65)
Big leaf maple	8,110	(18.86)
Canyon live oak	8,160	(18.98)
Chinkapin	8,320	(19.35)
Douglas-fir	8,760	(20.37)
Madrone	8,230	(19.14)
Ponderosa pine	9,020	(20.98)
Red alder	8,298	(19.30)
Tan oak	8,220	(19.12)
Western hemlock	8,550	(19.89)
Western red cedar	8,840	(20.56)
Bark		
Black oak	7,350	(17.09)
Chinkapin	9,630	(22.40)
Douglas-fir	9,430	(21.93)
Ponderosa pine	10,350	(24.07)
Red alder	8,360	(19.44)
Tan oak	8,315	(19.34)
Western hemlock	9,450	(21.98)
Western red cedar	8,960	(20.84)
Cull material		
Douglas-fir	8,270	(19.23)
Douglas-fir[a]	8,650	(20.12)
Madrone	8,390	(19.51)
Red alder	9,670	(22.49)
Tan oak	8,140	(18.93)

[a] Advanced brown and/or white rot.

TABLE VII. Bulk Density of Selected Biomass Fuels [a]

Fuel	Fuel lbs/ft³	(kg/m³)	Ash lbs/ft³	(kg/m³)
Agricultural				
Cherry pits	36.3	(580)	10.9	(175)
Coconut fiber and pith	5.5	(90)	–	
Coconut shell	33.3	(535)	–	
Gin trash	6.1	(95)	15.6	(250)
Grape pomace	37.7	(620)	–	
Grape vines	13.3	(215)	–	
Olive pits	34.0	(545)	–	
Peach pits	30.0	(480)	6.2	(99)
Peanut shells	14.2	(225)	23.4	(375)
Pistachio nuts	37.2	(595)	–	
Plum pits	35.1	(560)	–	
Rice hulls	18.1	(290)	9.9	(160)
Tomato pomace	13.4	(215)	10.6	(170)
Walnut shells	33.3	(535)	–	
Wood				
Black oak	23.0	(365)	–	
Big leaf maple	10.3	(165)	–	
Canyon live oak	19.7	(314)	–	
Chinkapin	12.6	(200)	–	
Douglas-fir	10.7	(170)	–	
Madrone	17.7	(280)	–	
Red alder	11.1	(175)	–	
Tan oak	19.8	(315)	–	

[a] Material minus 0.0787 inch (2 mm) and plus 0.0394 inch (1 mm).

TABLE VIII. Relationship Between Screen Fractionation and Bulk Density of Selected Agricultural Residues[a]

Sample	Size fractionation (%)	Bulk density lbs/ft^3 (kg/m^3)		Ash (%)
Orchard prunings				
plus 0.0787 inch (2 mm)	74.4	19.7	(315)	2.0
minus 0.0197 inch (500 μm) and plus 0.0787 inch (2 mm)	12.6	24.8	(395)	62.3
minus 0.0197 inch (500 μm)	13.0	68.3	(109)	90.8
Vineyard prunings				
plus 0.0787 inch (2 mm)	63.0	11.0	(175)	12.0
minus 0.0197 inch (500 μm) and plus 0.0787 inch (2 mm)	28.9	22.7	(360)	71.4
minus 0.0197 inch (500 μm)	8.1	69.9	(1120)	92.6
Gin trash				
plus 0.0787 inch (2 mm)	53.5	7.7	(125)	10.1
minus 0.0197 inch (500 μm) and plus 0.0787 inch (2 mm)	22.5	20.2	(325)	35.4
minus 0.0197 inch (500 μm)	24.0	29.1	(1165)	41.4

[a]Results are site specific only.

TABLE IX. Selected Ash Fusion Temperatures

Fuel	Initial °F (°C)	Softening °F (°C)	Hemispherical °F (°C)	Fluid °F (°C)
Vineyard prunings				
Dirt sample				
Oxidizing	2,290 (1254)	2,405 (1318)	> 2,700 (1482)	
Reducing	2,270 (1243)	2,340 (1282)	2,640 (1449)	> 2,700 (1482)
As received				
Oxidizing	2,220 (1216)	2,225 (1218)	2,235 (1224)	2,245 (1229)
Reducing	2,215 (1213)	2,220 (1216)	2,230 (1221)	2,240 (1227)
Clean chips				
Oxidizing	2,395 (1313)	2,495 (1368)	2,505 (1374)	2,595 (1424)
Reducing	2,390 (1310)	2,480 (1360)	2,500 (1371)	2,520 (1382)
Tan oak				
Wood sample				
Oxidizing	2,535 (1390)	2,625 (1440)	2,640 (1449)	2,655 (1457)
Reducing	2,510 (1377)	2,620 (1438)	2,635 (1446)	2,650 (1454)
Madrone				
Wood sample				
Oxidizing	2,320 (1271)	2,425 (1330)	2,465 (1352)	2,560 (1404)
Reducing	2,315 (1268)	2,370 (1299)	2,390 (1310)	2,410 (1321)

IV. DISCUSSION

A. Chemical Characterization of Biomass Fuels

The values obtained in the proximate analysis, ultimate analysis, and adiabatic calorimetry can be used as a means of comparing various biomass fuels. For purposes of comparison, representative values for coal are also considered as presented in Tables X and XI. To facilitate comparisons a summary table of the volatile-to-fixed-carbon ratio calculated from the ultimate analysis along with the higher heating values are presented in Table XII. A brief summary of Table XII indicating the overall ranges of values is presented in Table XIII.

The volatile to fixed carbon ratio is an indication of the volatility of a specific fuel sample. Coal has a low volatile to fixed carbon ratio ranging from 0.06 to 0.94 as a result of its high degree of aromaticity and its low oxygen content. The volatile-to-fixed-carbon ratio for agricultural and wood fuels ranges from 2.8 to 9.9 as shown in Table XII. As a result these materials are less aromatic, more oxygenated, and more reactive. A measure of this reactivity is the atomic ratio of hydrogen to carbon (H/C) and oxygen to carbon (O/C) calculated on an atomic basis.

These ratios indicated, for example, that the biomass fuels have 1.27 to 1.64 hydrogen atoms per carbon atom while coal has only 0.43-1.47 (ignoring lignite) hydrogen atoms per carbon atom, as summarized in Table XIII. Similarly biomass has 0.44 to 0.94 oxygen atom per carbon atom while coal has 0.04 to 0.06 oxygen atoms per carbon atom. Wood is very reactive, even at low temperatures wood can undergo alkaline hydrolysis, hydrogenolysis, nitration, and chlorination. Because of the hydroxyl groups of cellulose it behaves like a polyalcohol readily undergoing esterification by nitration, acetylation, etherification, substitution reactions with halogens, nitrogen dioxide, and ammonia, carboxylation, and formation of aldehydes. For example, using sulfuric acid Klason lignin and low molecular weight sugars are formed from wood while coal remains unreacted under similar conditions.

As shown in Tables XII and XIII, there is a large range in the volatile-to-fixed carbon, H/C, O/C, and the higher heating values (HHV). It can be seen that bark has a lower volatile-to-fixed carbon ratio than wood. Similarly, the hardwoods have a slightly higher volatile-to-fixed carbon ratio than the softwoods, while the cull material has the highest volatile to fixed carbon ratio with the greatest range.

TABLE X. *Proximate Analysis, Volatile-to-Fixed Carbon Ratio and Heat Contents of Selected Coals (Baumeister et al., 1983)*

Type	Volatile matter (%)	Fixed carbon (%)	Ash (%)	Volatile-to-Fixed carbon ratio	Higher heating value Btu/lb (MJ/kg)
Anthracite	5.4	85.0	9.6	0.06	12,880 (29.98)
Semianthracite	10.9	81.6	7.5	0.13	13,880 (32.31)
Low-volatile bituminous	18.3	76.3	5.4	0.23	14,400 (33.52)
Medium-volatile bituminous	25.0	68.9	6.1	0.36	14,310 (33.31)
High-volatile bituminous	39.9	51.6	8.5	0.77	12,170 (28.30)
Subbituminous A	42.2	54.2	3.6	0.78	10,650 (24.80)
Subbituminous B	43.2	52.5	4.3	0.82	9,610 (22.38)
Subbituminous C	41.6	51.6	6.8	0.81	8,500 (19.79)
Lignite	45.7	48.4	5.9	0.94	7,000 (16.30)

TABLE XI. *Ultimate Analysis, Hydrogen-to-Carbon Ratio and the Oxygen-to-Carbon Ratio of Selected Coals (Baumeister et al., 1983)*

Type	Carbon (%)	Hydrogen (%)	Oxygen (%)	Nitrogen (%)	Sulfur (%)	Ash (%)	H/C	O/C
Anthracite	79.7	2.9	6.1	0.9	0.8	9.6	0.43	0.06
Semianthracite	81.4	3.8	4.0	1.6	1.7	7.5	0.56	0.04
Low-volatile bituminous	83.2	4.6	4.7	1.3	0.8	5.4	0.66	0.04
Medium-volatile bituminous	81.6	5.0	4.9	1.4	1.0	6.1	0.73	0.05
High-volatile bituminous	67.7	5.6	14.4	3.9	2.5	8.5	1.00	0.16
Subbituminous A	60.4	6.0	27.4	1.2	1.4	3.6	1.19	0.34
Subbituminous B	53.9	6.9	33.4	1.0	0.5	4.3	1.53	0.46
Subbituminous C	50.5	6.2	35.5	0.7	0.3	6.8	1.47	0.53
Lignite	40.6	6.9	45.1	0.6	0.9	5.9	2.04	0.83

TABLE XII. Selected Fuel Properties Including Coal

Fuel	Volatile/fixed carbon ratio	Hydrogen/carbon ratio	Oxygen/carbon ratio	Higher heating value Btu/lb (MJ/kg)
Coal				
Anthracite	0.06-0.13	0.43-0.56	0.04-0.06	12,880 (29.98)-13,880 (32.31)
Bituminous	0.23-0.77	0.66-1.00	0.04-0.16	12,170 (28.30)-14,400 (33.52)
Subbituminous	0.78-0.82	1.19-1.47	0.34-0.53	8,500 (19.79)-10,650 (24.80)
Lignite	0.94	2.04	0.83	7,000 (16.30)
Agriculture				
Pits and nuts	4.0-5.7	1.55	0.66	8,350 (19.42)-9,350 (21.75)
Hulls, trash, and shells	4.0-4.6	1.36	0.69	6,400 (14.89)-8,840 (20.56)

Pomace	3.5–7.9	1.27	0.44	9,373 (21.81)–10,220 (23.77)
Prunings	5.2–5.7	1.41–1.45	0.62–0.66	7,220 (16.79)–8,190 (19.05)
Wood				
Softwood	6.6–6.9	1.41	0.60	8,550 (19.89)–9,020 (20.98)
Hardwood	5.5–7.8	1.43–1.52	0.65–0.72	8,020 (18.65)–8,320 (19.35)
Cull	5.5–9.9	1.40–1.48	0.60–0.69	8,270 (19.23)–9,670 (22.49)
Bark				
Softwood	2.8–3.7	1.18–1.28[a]	0.49–0.61[a]	8,960 (20.84)–10,350 (24.07)
Hardwood	3.7–4.8	1.25–1.64[a]	0.53–0.94[a]	7,350 (17.09)–9,630 (22.40)

[a] Postulated from literature.

TABLE XIII. Summary of Selected Fuel Properties

Fuel	Volatile/fixed carbon ratio	Hydrogen/carbon ratio	Oxygen/carbon ratio	Higher heating values Btu/lb (MJ/kg)
Coal[a]	0.06-0.82	0.43-1.47	0.04-0.53	8,500 (19.79)-14,400 (33.52)
Agricultural	3.50-7.90	1.27-1.55	0.44-0.69	6,400 (14.89)-10,220 (23.77)
Wood	5.50-7.80	1.41-1.52	0.60-0.72	8,020 (18.65)-9,020 (20.98)
Cull material	5.50-9.90	1.40-1.48	0.60-0.69	8,270 (19.23)-9,670 (22.49)
Bark	2.80-4.80	1.18-1.64	0.49-0.94	7,350 (17.09)-10,350 (24.07)

[a] Does not include lignite.

This can be understood by the chemical composition of wood fuels. The major components of wood are the extractives, lignin, cellulose, and hemicelluloses. The hemicellulose and cellulose fractions are called the holocelluloses or total carbohydrates. Bark has phenolic acids in addition to the components usually found in wood. The phenolic acid fraction is similar to tannins except that it cannot be removed by solvent extraction because of its molecular size. The chemical composition of selected wood and bark fuels is presented in Table XIV. The carbohydrate and lignin fractions have significantly different compositions. This difference is manifested in the amount of char and volatiles formed as indicated below (Shafizadeh, 1982).

Type	HHV	% Char	% Volatiles
Cellulose	7,460 (17.47)	14.9	85.1
Lignin	11,470 (26.86)	59.0	41.0

The carbohydrate fraction of bark is much less than that of wood as can be seen in Table XIV. However, bark has less lignin than wood but has a high phenolic acid content. These phenolic acids are more complex than lignin and as a result less oxygenated and would thus have a higher heat content and form more char. Thus, it becomes obvious why softwood bark and in most cases hardwood bark have a lower volatile-to-fixed carbon ratio, a lower H/C and O/C ratio, and a higher HHV.

Cull material has a much broader range of volatile-to-fixed carbon ratio, H/C and O/C ratio. The low values of the volatile to fixed carbon ratio can in most cases be attributed to white rot and/or pocket rot in which case cellulose is degraded increasing the amount of lignin present and as a result increasing the HHV and decreasing the H/C and O/C ratio. For brown rot in which lignin is degraded, the cellulose content is increased on a volume basis and as a result the volatile-to-fixed carbon ratio increases as well as the H/C and O/C ratio while the HHV decreases.

For hardwoods the volatile-to-fixed carbon ratio increases as well as the H/C and O/C ratio, while the HHV decreases because these woods contain less lignin and more cellulose than softwoods. This type of discussion can also be applied to other types of wood. Although no experimental results are available at this time, the following general results are expected when compared to softwood.

TABLE XIV. Chemical Composition of Selected Wood and Bark Fuels[a,b]

Species	Lignin (%)	Phenolic acid[a][c] (%)	Carbohydrate (%)
Wood			
Balsam fir	30.2	N/A	69.8
Common beach	25.4	N/A	74.6
Eastern hemlock	31.7	N/A	68.3
Norway spruce	28.1	N/A	71.9
Paper birch	22.5	N/A	77.5
Red gum	32.8	N/A	67.2
Bark			
Balsam fir	19.1	16.2	64.7
Black spruce	19.2	14.6	66.2
Eastern hemlock	27.2	21.3	51.5
Jack pine	18.8	36.2	45.0
Lodgepole pine	10.0	17.3	72.7
Slash pine	32.1	29.2	38.7
Sugar pine	27.8	41.3	30.9

[a] Adopted from Sjostrom, 1981, and Sarkanen and Ludwig, 1971.
[b] Values corrected to add to 100 percent.
[c] No phenolic acid present in wood.

Type	Volatile-to-fixed carbon	H/C	O/C	HHV
Juvenile wood	greater	greater	greater	lower
Tension wood	greater	greater	greater	lower
Compression wood	lower	lower	lower	greater
Earlywood	greater	greater	greater	lower
Latewood	lower	lower	lower	greater

Similar results for agricultural residues are not available at this time. As can be seen by the large range in the volatile-to-fixed carbon ratio, the H/C and O/C ratios, and the HHV, there must be very distinct differences in chemical composition even though the overall range is similar to that of bark and wood. In general the pomaces are significantly different than the other agricultural residues on both the low and high range of values.

B. Existing Combustion Facilities

There are many different sources of agricultural residues currently being produced, the majority of which are currently being field burned or disked back into the soil. These residues can be divided into two general categories, those residues generated in the field and those residues generated at food and other processing facilities. Field residues include straws and orchard and vineyard prunings. The residues currently being generated at processing facilities can be further divided into a low and a high moisture content fraction. The low moisture content processing wastes include pits, shells, rice hulls, and cotton gin trash while the high moisture content fraction including such things as grape and tomato pomace and sugar beet residues. Of the residues the largest single fraction currently being combusted, in for example, the state of California, is the low moisture content fraction. This is true for several reasons, particularly in the case of the pits and shells, as indicated below.

(1) They are readily available at the processing facilities;

(2) Their cost of collection can be attributed to the processing operation; and

(3) They are readily combusted in conventional combustion devices.

A list of facilities in the state of California currently burning pits, shells, rice hulls, and cotton gin trash is presented in Table XV. These facilities produce process heat, steam, and cogenerate electricity. Rice hulls are being burned to offset the use of natural gas in cement kilns. Cotton gin trash is being used to offset the use of natural gas in drying the cotton before it is shipped to the gin.

V. CONCLUSIONS

This chapter has attempted to present the chemical and physical properties of wood and agricultural residues important to fuel behavior. There is a significant potential for agricultural residues in reducing fossil fuel consumption in the food processing industry as well as use in stand-alone power plants. However, there are certain environmental related problems that must be overcome before these fuels become widely accepted as previously indicated.

To facilitate the use of agricultural as well as wood fuels much research is needed. A list of specific research needs for both agricultural and wood residues is listed below.

(1) Establishing a data base of test results by residue type and region of the United States. This data base should have some degree of statistical significance.

(2) Develop standard test procedures similar to ASTM standards for coal. This is needed because of the high volatile-to-fixed carbon ratio and the low specific gravity of biomass residues.

(3) Perform micro-analysis of biomass fuels to determine where the ash, nitrogen, and sulfur are concentrated.

(4) Provide testing for the difference between the inherent and entrained ash content of the material as a means of choosing optimal collection and storage methods.

(5) Use the lignin and holocellulose determination to relate the lignin to holocellulose ratio to the volatile to fixed carbon ratio and H/C and O/C ratio for both agricultural and wood residues.

(6) Perform elemental analysis of the ash to facilitate compliance with federal and state regulations for solid waste disposal.

(7) Perform experimental combustion of the various residues to determine the amount of nitrogen and sulfur converted to oxides of nitrogen and oxides of sulfur, respectively, as a function of combustion methods and conditions. This could involve testing existing commercial units.

TABLE XV. List of Facilities Currently Combusting Pits, Shells, Rice Hulls, and Cotton Gin Trash[a,b]

Project	Type	Energy use	Fuel type	O.D. Tons fuel consumed (Year)
Calaveras Cement	Suspension system	Process heat	Rice hulls	20,000
Diamond Sunsweet	Spreader stoker	Cogeneration 4.5 MW	Walnut shells	41,200
Farmers Coop. Gin	Suspension system	Cogeneration 1.7 MW	Cotton gin trash	3,670
Kerman Gin	Tunnel burner	Process heat	Cotton gin trash	1,090
Lindsay Olive	Fluidized bed	Process steam	Olive pits	5,030
Port Costa Clay Products[a]	Suspension system	Process heat	Rice hulls	17,170
Tri-Valley Growers	Pile burner	Cogeneration 4.5 MW	Pits and shells Woodwaste	10,000 25,000
Westside Farmers Coop.	Tunnel burner	Process heat	Cotton gin trash	870

[a] Startup second quarter of 1984.
[b] Data from the California Energy Commission.

(8) Determine the applicability of the forms of sulfur analysis to biomass fuels as a means of predicting sulfur oxide emission levels.

(9) Perform thermogravametric analysis as a tool in determining the rate of weight loss of a fuel under various combinations of reducing and oxidizing atmospheres.

(10) Perform differential scanning calorimetry as a means of measuring the rate of heat release under various combinations of reducing and oxidizing atmospheres.

(11) Establish computer models using data from the thermogravometric, differential scanning calorimeter, and other tests to predict heat release rates, chemical reactions in the fuel bed, and the emissions formed.

These types of research are necessary to promote the use of agricultural and wood residues. A sound data base is required before agricultural residues can be effectively burned in direct combustion and/or gasification systems. The same tests are also needed for forest and forest product mill residues. The majority of these tests were done long ago and as a result changes in the resource base are not reflected in the limited data available. Even though biomass has been burned for generations in industrial combustors, the need today has changed in light of return on investment and environmental concerns. As a result, much research is needed.

REFERENCES

Baumeister, T. et al. 1981. "Mark's Standard Handbook for Mechanical Engineers." 8th ed. McGraw Hill Book Co., New York.

Browning, B. L. 1967. "Methods of Wood Chemistry," Vols. 1 and 2. Interscience Publishers, New York.

Mingle, J. G., and Boubel, R. W. 1968. Proximate Analysis of Some Western Wood and Bark. *Wood Science* 1(1):29-36.

Sarkanen, K. V., and Ludwig, C. H. 1971. "Lignins Occurrence, Formation, Structure, and Reactions." Wiley-Interscience, New York.

Shafizadeh, F. 1982. Chemistry of Pyrolysis and Combustion of Wood. *In* "Progress in Biomass Conversion, Vol. 3" (K. V. Sarkanen, D. A. Tillman, and E. J. Jahn, ed.). Academic Press, New York.

Sjostrom, E. 1981. "Wood Chemistry Fundamentals and Applications." Academic Press, New York.

FACTORS INFLUENCING DILUTE SULFURIC ACID PREHYDROLYSIS OF SOUTHERN RED OAK WOOD

J. F. Harris
R. W. Scott
E. L. Springer
T. H. Wegner

U.S. Department of Agriculture
Forest Service
Forest Products Laboratory
Madison, Wisconsin

I.	INTRODUCTION	102
II.	PRINCIPLES FOR DATA INTERPRETATION	103
	A. Xylose Removal Rate	103
	B. Xylose Decomposition Rate; Calculation of Xylose Yields	106
	C. Calculation of Furfural Yields	107
	D. Acid-Neutralizing Capacity of Wood	108
	E. Effective Acidity of the Hydrolysis Solution	111
III.	SMALL-SCALE HYDROLYSIS STUDIES	113
	A. Effect of Acidity and Temperature at Low pH	113
	B. Comparison of Water and Dilute Acid Hydrolysis	115

[This paper is in the public domain.]

IV. HYDROLYSIS OF OAK CHIPS IN DIRECT STEAM 118

    A. Chip Preparation 118

    B. Digester Operation 119

    C. Water Movement 123

    D. Effective Acid Strength 128

    E. Product Yields 129

VIII. SUMMARY 138

I. INTRODUCTION

The basic objective of this work was to investigate the fundamental principles involved in hydrogen-ion catalyzed prehydrolysis. To this end the yields of soluble carbohydrates and furfural were related to process variables: (1) reaction time, (2) temperature, (3) acid concentration (pH), and (4) quantity of hydrolyzing solution. The examination is restricted, almost completely, to experience gained at the Forest Products Laboratory (Madison, Wisconsin), where recent studies have been part of a larger effort to evaluate the production of ethanol from hardwoods (Harris et al., 1983). We include discussions of data obtained from hydrolyses of finely divided wood in small glass ampoules and from hydrolyses of wood chips in a laboratory digester. Most of the reported data are on southern red oak, but the conduct and interpretation of the hydrolyses that supplied these data should be applicable to other hardwoods and to other plant material.

In this chapter we use the term, prehydrolysis, to indicate dilute acid hydrolysis which fractionates plant cell wall material into water-soluble hemicellulose fragments and water-insoluable lignocellulose. The definition implies an ideal chemical fractionation without regard for preserving fibers. During dilute acid prehydrolysis at pH 1 to 2, plant materials will often release 25 to 35% of their original mass in a few minutes. The major hardwood component susceptible to prehydrolysis is the partially acetylated 4-O-methylglucuronoxylan, subsequently referred to here as xylan. The major xylan constituent, the xylose anhydride monomeric unit in the backbone of the polymer, will be referred to as xylose anhydride or simply as xylose.

II. PRINCIPLES FOR DATA INTERPRETATION

Removal of constituents from wood by acid hydrolysis involves complex physical changes and chemical reactions which are related to the complexity of the wood itself. In spite of this, certain mathematical expressions related to more homogeneous systems are useful. For example, we were able to predict hydrolytic yields of xylose and furfural from previous data and from new data reported here. The necessary equations appear in the following discussions of xylose removal, xylose decomposition and yield, and furfural production. Application of these equations required a knowledge of the acidity of the hydrolyzing solution which is also discussed.

A. Xylose Removal Rate

The hydrolytic removal of xylose from hardwood lignocellulose under uniform conditions of temperature, acidity, and liquid-to-solid ratio (L/S) is typically represented by a curve on a semilogarithmic plot such as that in Figure 1. The first portion of this curve, to about 40% remaining, is quite linear. Thereafter, the removal rate continuously decreases to a minimum value equal to the hydrolysis rate of cellulose at the particular conditions employed. This is deduced from the fact that on prolonged hydrolysis the composition of the residual carbohydrate reaches a constant value (Harris et al., 1963), indicating that a small portion (about 1%) of the original potential xylose is intimately associated with the resistant cellulose. This final constant removal rate is well beyond the range shown in Figure 1. For that portion of the curve in which the slope is decreasing, neither the position nor the slope at any point can be predicted. Consequently it is necessary to collect experimental data in the region of interest. This region for a prehydrolysis aimed at maximum xylose recovery, is a residue containing between 5 and 10% of the original potential xylose of the wood.

Throughout the discussions that follow, the term "rate constant" refers to the absolute value of the slope of a removal curve such as that in Figure 1. It should be understood that this rate constant is dependent on, but not directly related to, the rate constants for individual bond cleavages.

Springer (1966) investigated the effect of acid concentration and temperature on the initial rate constant for xylose removal from aspen wood and found that it could be described by the equation:

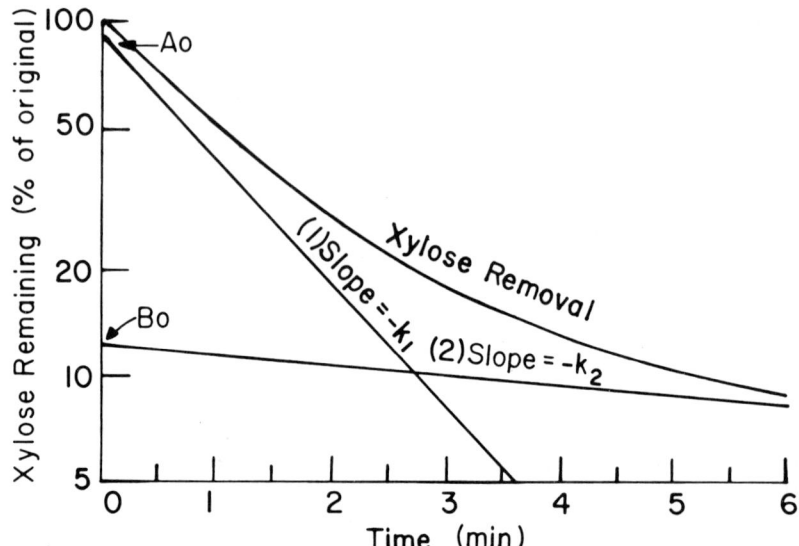

FIGURE 1. Empirical hydrolytic removal curve of xylose from wood. Percentage of original wood xylose remaining in the residue. (1) and (2) are linear components of the xylose removal curve as in Eq. (2).

$$\log_{10}(K/CH) = 15.083 - 6171.3/T + 0.22219 CH \quad (1)$$

Where

T = absolute temperature, °K

CH = molarity of hydrogen ion, $[H^+]$

K = rate constant, min^{-1}

Springer and Zoch (1968) also compared such initial removal rate constants for several wood species and found only small variations. These corresponded with differences in the proportion of 4-0-methylglucuronic acid side chains in the xylans.

An analytical expression for the xylose removal rate allows one to conveniently calculate xylose and furfural yields, and to predict shifts of the curve when minor temperature and acidity changes are made. The derivation of the

Dilute Sulfuric Acid Prehydrolysis

functional form used here is described below and illustrated in Figure 1.

Assume the removal curve to be the sum of two linear functions: the two lines plotted on the semilogarithmic scale of Figure 1.

XR = percentage of original xylose remaining in the residue,

$$= A \ \exp(-k_1 t) + B_o \exp(-k_2 t), \text{ where } t = \text{time, minutes} \quad (2)$$

Because the plot is presented as percentage removal, at $t = 0$,

$$XR = A_o + B_o = 100; \quad (3)$$

so that

$$XR = A_o \exp(-k_1 t) + (100 - A_o)\exp(-k_2 t) \quad (4)$$

An experimental data set can be used with Equation (4) to obtain the values of A_o, k_1, and k_2 for the best fit. These values substituted into Equation (2) yield the desired analytical expression.

In the preceding discussion it is assumed that the removal curve of Figure 1 was obtained under conditions of constant temperature, L/S, and hydrogen ion concentration, the independent variables determining removal rates. It is thought that the L/S has no significant effect on the removal rate for ratios 1.0 to 5.0. To estimate the effect of minor changes in acidity and temperature, it can be assumed that k_1 and k_2 are proportionally affected by the changes in the same manner as K in Quation (1). This will be valid only over a very limited range. In general, experimental data must be collected for a particular species at the temperatures, acidities, and times of interest.

One should not be tempted to attach fundamental significance to the rate constants, k_1 and k_2, in Equation (4). It is apparent from Figure 1 that the values of the parameters will depend on the particular range in which data are collected. The admonition concerning k_1 and k_2 is warranted because they have been cited as evidence of two different xylan polymers (Lee and McCaskey, 1982). However, the dependence on the particular selection of data makes a purely chemical interpretation suspect.

The reasons for a decreasing removal rate of xylose are open to speculation. Dissolution depends not only on the rates of hydrolysis of $\beta(1\rightarrow 4)$ bonds in xylan, but also on other variables: (1) solubilities of various oligomers, (2) changing chemical structure of xylan as acetyl and uronic acid ratios to xylose change, (3) increasing proportion of resistant bonds, (4) changing diffusion characteristics as the basic lignocellulosic structure is attacked by acid, and (5) the distribution of acid throughout the wood. This last is perhaps quite important. It is known that not all of the water in the wood structure is accessible to acid. Sookne and Harris (1940) observed that 0.12g of water per gram of cotton cellulose was inaccessible to HCL. We observed that oak wood, which was soaked in 0.73% H_2SO_4 for several days, excluded 10% of the acid calculated to be taken up on the basis of water uptake.

B. *Xylose Decomposition Rate; Calculations of Xylose Yields*

Data on the decomposition rate of xylose, covering a broad range of temperature, acidity (H_2SO_4), and concentration, are available (Root, 1956). The first order rate constant is correlated by the expression:

$$k = 2.72 \, \alpha\delta\gamma(CA)\exp[-35.7(473.1-T)(1/T)] \quad (5)$$

Where

T = absolute temperature, °K

CA = acid normality

α = $\alpha(CX)$, a function of xylose molarity, CX

δ = $\delta(T)$, a function of temperature

γ = $\gamma(T,CA)$, a function of temperature and acid normality

No functional forms are available for α, δ, and γ, but their relationship to the independent variables is given in tabular form in the original work.

Root's data, along with xylose removal-rate data, were used to calculate the xylose content of prehydrolysate solutions as a function of time. Numerical integration was required because there is no mathematical expression for k, which varies with xylose concentration as the reaction proceeds. Because xylose is liberated during hydrolysis, the value of CX varies, changing the values of α. The change in CX also indirectly affects the acid normality since the volume of the solution changes. Thus γ also varies throughout the reaction.

The calculation of xylose yields, from (1) the xylose removal curve, Equation (4) and (2) Root's data, Equation (5), assumes that the xylan is released into solution as monomeric xylose rather than in oligosaccharides. Because decomposition of a sugar molecule involves its reducing end group (Feather and Harris, 1973), the protection afforded xylose by combination in oligomers would decrease its loss. Experimentally, we found that in dilute sulfuric acid solutions at maximum xylose yields, up to 30% (often 10 to 20%) of the total xylose content of the solution was present in oligomeric form.

C. Calculation of Furfural Yields

Furfural production from xylose is illustrated in the following mechanism:

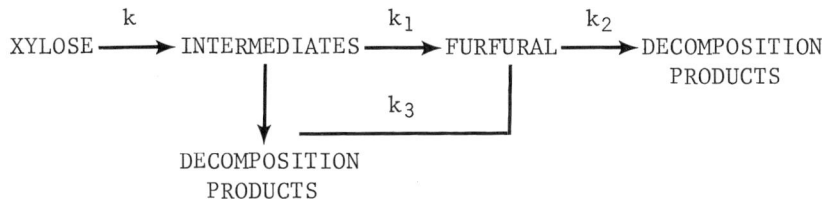

This mechanism is a simplified model for the complex group of reactions associated with the acid-catalyzed decomposition of xylose. There are at least three intermediates, all of which probably react with furfural at different rates. The decomposition of furfural, reacting alone in solution, is not first order, and it is likely that its decomposition products are also participants. Nevertheless, this mechanistic scheme was successfully applied by Root (1956) to correlate his extensive set of data.

An important point that may be deduced from the mechanism is that furfural is produced very efficiently during prehydrolysis. In the early stages of the reaction, the concentration of furfural is low, and the loss rate is low since it depends on the furfural concentration. The yield of furfural based on xylose reacted is high, beginning at 100% but decreasing continuously as the amount of xylose reacted increases.

The differential production of furfural was approximated by Root (1956) using the equation:

$$\frac{dF}{dt} = k[X - aF - bF(X/X_0)] \tag{6}$$

Where

X and F = molar concentrations of xylose and furfural

X_0 = initial molar concentration of the xylose solution

k = reaction rate constant for xylose disappearance

a = $a(T, X_0)$ (see Root, 1956)

$b = b(T, X_0)$ (see Root, 1956)

T = absolute temperature, °K

The functions, a and b, evaluated from experimental data, resulted in an excellent correlation. An integrated solution of the above equation was also presented, but only the differential form is used here.

Root's experimental procedure was to load small ampoules with solutions of various acid and xylose concentrations, react them for timed intervals, and measure the furfural yield. The quantity X_0 was a clearly defined, experimental variable which was included in the empirical correlation. However, in applying the correlation to prehydrolysis, a problem arose over the value to use for X_0. Root's correlation was based on experiments in which the xylose concentrations, initially X_0, decreased exponentially with time. In prehydrolysis, the xylose concentration is initially zero, rises to a maximum, and then decreases. During prehydrolysis the concentration relationships between the intermediates, products, and xylose are obviously very different from that present in the original study.

Furfural production in prehydrolysis was calculated by simultaneous numerical integration of the xylan removal rate curve, Equation (4), the xylose degradation rate equation, Equation (5), and the furfural rate equation, Equation (6). The furfural production rate, dF/dt, was evaluated at each time increment by assuming two values for X_0, (1) the actual xylose concentration at the moment and (2) the hypothetical xylose concentration assuming all the xylose released to be present. After all combinations of these two values were used to evaluate dF/dt, the minimum was selected. Calculated furfural yields are compared with experimental values in a later section.

D. Acid-Neutralizing Capacity of Wood

The catalytic acid concentration, or pH, of the hydrolyzing solution affects reaction rates in both the solid and liquid phase and is an important process variable. Its value depends upon the quantity and type of applied acid, the L/S, and the neutralizing capacity of the substrate. If large quantities of acid are used, this latter factor is not important but at conditions of commercial interest where low L/S and dilute acid solutions are used its effect is significant. In our investigations as much as 40% of the applied acid was neutralized by the wood and was unavailable as a catalyst.

Due to the presence of weakly acidic organics the pH of green wood varies from 4 to 6 depending on the species. Acetyl groups are released as acetic acid when wood is heated, thereby decreasing the pH below that of the green wood. However, acetic acid and other weak organic acids do not contribute catalytic hydrogen ions at pH 1 to 1.5 because they are not dissociated at this high acidity. The principal components that are effective in decreasing the acidity are salts of 4-0-methylglucuronic acid. Because the pK of this acid is about 3, at the pH levels of hydrolysis ($\sim$1.5) the organic salt is completely converted to the undissociated acid. Thus each equivalent of salt neutralizes an equivalent of acid. Not all of the 4-0-methylglucuronate groups are present in the wood as salts, some are esterified. The measurements, described below, on the southern red oak used in this study indicate that 64% are in salt form. It should be noted that, in contrast to neutralization of uronic acid salts, the hydrolysis of uronic acid esters does not consume acid. Calcium oxalate is a minor but frequent constituent of wood (Krasowski and Marton, 1982). It will contribute to the neutralizing capacity if present. However, only one of the carboxyl groups will be effective at the conditions of prehydrolysis. There are some inorganics in wood which do not contribute to its neutralizing capacity, these are primarily silica and minor quantities of inorganic salts.

Direct titration of wood extracts to determine the capacity of wood to neutralize strong acids is complicated by interference from organic acids and colored extractives. We were unsuccessful in this direct approach and used an alternate method employing the titration of the ash described below. However, Veeraraghavan et al. (1982) recorded the changing pH from the eluate of a percolation reactor and estimated the neutralizing capacity of red oak.

We define the "neutralizing capacity" of wood to be the milliequivalents (meq) of carboxylate ions in 1 kg of oven-dried wood. When wood is ashed, the carboxylate-combined metal ions are converted to carbonates, which are usually the major components of wood ash. The meq of metal ion carbonates in the ash from 1 kg of wood is about equal to the neutralizing capacity. A possible discrepancy caused by trivial amounts of oxalate was considered negligible. We determined the neutralizing capacity of wood by titration of the alkalinity of wood ash (Scott et al., in press). An excess of standard sulfuric acid was added to the ash from a weighed sample of wood, and the excess acid was back-titrated with standard sodium hydroxide to a phenolphthalein end point. Corroboration of this method was obtained by quantitative atomic emission (and absorption) flame spectrophotometry for Ca, Mg, K, and Na, and calculation of their neutralizing capacity. For a southern red oak sample the average titrated value was 118 meq/kg and the average sum of the four cations was 123 meq/kg.

Representative samples of wood must be used for ashing since there can be differences between logs, partly because of different amounts of ash in heartwood than in sapwood (Tout et al., 1976). We found much less ash in heartwood than in sapwood of southern red oak.

The effect of the neutralizing capacity of the wood on the xylose removal rate was investigated by comparing the hydrolysis rates of two samples of the same wood with different ash contents (Springer, unpublished). Both samples were of finely divided (40 mesh) southern red oak heartwood (Table I, sample sample 1). From the original wood a de-ashed wood sample was prepared by washing with dilute hydrochloric acid (H_2SO_4 was used in the hydrolysis). The hydrolysis rates of the original wood and the de-ashed material were compared. Kinetic studies were done in 5 mm glass ampoules at a temperature of a 120°C and a L/S of 3.0. The low temperature was chosen to obtain the necessary experimental accuracy.

As previously discussed, the initial xylose removal, down to ~40% remaining xylose, is linear on a semilogarithmic plot. Over this removal range the xylose loss can be represented by a kinetic constant, k, similar to a first order reaction rate constant. This constant is almost directly proportional to acid concentration when the solution is dilute (Equation 1).

Initial rate constants for the de-ashed wood were determined at various sulfuric acid concentrations. The result is shown in Figure 2. Other ampoules containing original wood and 0.20% H_2SO_4 were reacted at 120°C for various times. The resulting rate constant is given by the point in Figure 2. The horizontal displacement of this point from the line for de-ashed wood represents the loss in acid resulting from the neutralizing capacity of the wood. The effective neutralizing capacity of the original heartwood, as calculated from the acid loss (displacement of the point) is 31 meq/kg wood. For reasons presented in the following section, H_2SO_4 is assumed to have an equivalent weight of 98 in this calculation. From the cation composition of the ash, it was calculated that the original wood should have a neutralizing capacity of 37 meq/kg, which is in reasonable agreement with the value measured kinetically. This infers that a good approximation to acid strength can be made after determination of the neutralizing power of the ash. However, this is concluded only from hydrolyses of finely divided wood in small ampoules where most ash constituents were effectively neutralized, unlike hydrolyses with wood chips. It should be pointed out that the experimental conditions were limited to the 50% xylose removal range.

TABLE I. Analyses of Southern Red Oak (Quercus falcata Michx.) Wood Samples

Anhydride	Sample		
	1^a (%)	2^b (%)	3^b (%)
Glucose	40.3	37.8	41.8
Mannose	2.9	2.1	2.2
Xylose	19.3	18.4	18.7
Galactose		1.1	0.8
Arabinose		0.7	0.8
Uronic	2.9	3.3	2.7
Acetyl	3.7	4.3	3.5
Lignin	21.8	21.9	19.5
Ash	0.24	0.72	0.29
Extractives	5.5	6.7^c	6.6
Sum	96.6	97.0	96.9

[a] This sample was entirely heartwood. The paper chromatographic procedure used did not separate galactose from glucose nor arabinose from mannose.

[b] Mixed heartwood-sapwood samples analyzed spectrophotometrically except for galactose and arabinose by paper chromatography.

[c] This extract contains both hot water and benzene-ethanol extractives. Other extracts contained only benzene-ethanol extractives.

E. Effective Acidity of the Hydrolysis Solution

Previous investigators (Root, 1956 and Springer, 1966) have satisfactorily correlated the kinetic constants of prehydrolysis using hydrogen ion concentration [H^+], rather than the more fundamental hydrogen ion activity. We continue this practice and use the concentration of hydrogen ion as the measure of the catalytic acid activity. A precise calculation of [H^+] at reaction conditions, straightforward in principle, is not possible because of the lack of information on the various ionic equilibria involved. A suitable estimate is obtained by the simple procedure of assuming H_2SO_4 to be monobasic, accounting for the neutralizing capacity of the substrate and ignoring the possible precipitation of $CaSO_4$.

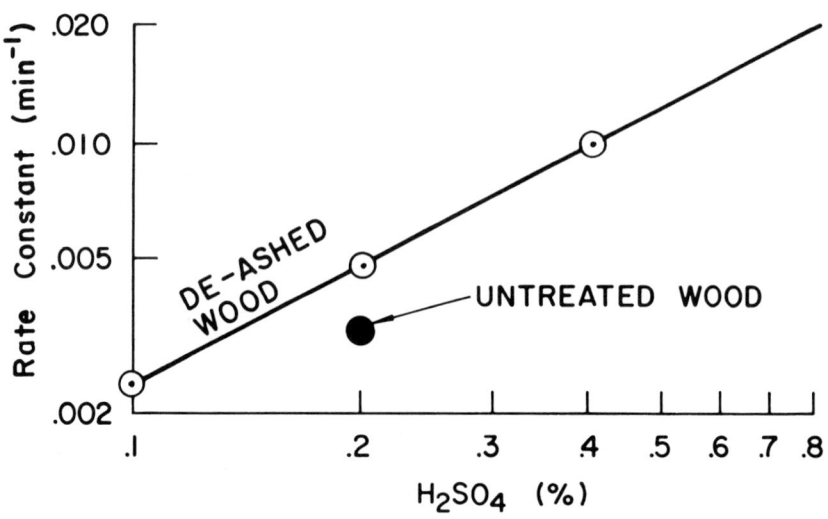

FIGURE 2. Initial hydrolysis rate constants as a function of applied acid concentration. Southern red oak, 120°C, liquid-to-solid ratio of 3.

For purposes of calculating the pH of H_2SO_4 solutions below pH 1.5 it can be assumed that the $[H^+]$ is supplied only by the first ionization of H_2SO_4. This assumption is based on an ionization constant of 0.0037 for HSO_4^- at 60°C and the fact that it decreases as temperature increases (Harned and Owen, 1958). At pH 1.4 and 60°C the HSO_4^- should be less than 10% ionized and at pH 1.0, less than 4% ionized. Since the ionization constant decreases approximately 20-fold if the temperature is raised to 170°C the secondary ionization can be ignored. As a consequence the gram equivalent weight of H_2SO_4 is 98 for determining $[H^+]$ and for the purpose of calculating acid loss due to the neutralization expected prior to and during hydrolysis.

The data presented previously demonstrated that, for southern red oak, the alkalinity of the ash corresponded to the cation content of the wood, and this in turn had a straight forward relationship to the rate of xylose removal from the wood. The effective acidity of the hydrolyzing solution was that of the charged solution reduced by an amount equal to the neutralizing capacity of the wood. In these kinetic experiments it was experimentally determined that nearly all of the cations in the finely divided wood were removed. In the case of chips not all of the neutralizing capacity is effective; it was found that 10 to 25% of the cations originally in the charged chips remained in the hydrolyzed residue. The neutralizing capacity for both charge and residue was assumed equal to that of its ash. The difference between these quantities is subtracted from the amount of acid charged to obtain the effective acid concentration $[H^+]$.

Because calcium is a major cation of wood ash, a further consideration in the calculation of acidity is the possible precipitation of $CaSO_4$ during hydrolysis with H_2SO_4. The acidity is affected because of the relative effectiveness of the calcium ion. In solution, where HSO_4^- predominates, each mol of Ca^{++} neutralizes two mols of H_2SO_4, in the solid phase only one. Calculation indicates that at ambient temperature much of the calcium would precipitate. However, as previously explained, at conditions of pH 1.0 to 1.5 and 170°C the concentration of $SO_4^=$ is very much less than that at ambient temperature. From this it was concluded that calcium sulfate would not precipitate at conditions of prehydrolysis.

III. SMALL-SCALE HYDROLYSIS STUDIES

A. Effect of Acidity and Temperature at Low pH

Small scale laboratory work on wood hydrolysis provides advantages: simplicity, rapid collection of data, control of pH and temperature, and minimal uncertainties due to diffusion. Baker and Krcmar (1956) heated sealed, 16 mm glass tubes containing 30-mesh sawdust of southern red oak for the measurements of residue xylose. The xylose yields and losses and the time for maximum yield were calculated using their data and Equation (5). Calculated maximum yields (Table II) were all above 80% except for the condition 190°C, 0.4% H_2SO_4. Reexamining this set of data and considering the experimental procedure led to the conclusion that this point was unreliable. From the remaining data, it was concluded that yields in excess of 80% would be obtained, and that increasing temperature and acidity both tended to increase maximum yields. In these experiments, data were collected on the residue composition only; no analysis was made on the accompanying hydrolysates, therefore it was not possible to confirm these calculated values.

TABLE II. *Maximum Xylose Yields Calculated from Residue Compositions. Hydrolysis of Southern Red Oak Sawdust.*

Temperature[a] (°C)	Added acid[a] (%)	Effective acid[b] (%)	Time at maximum yield (min)	Percentage of wood xylose		
				In residue[a] (%)	Maximum yield[b] (%)	Degraded[b,c] (%)
170	0.1	0.064	19.8	13.0	81.3	5.7
	0.4	0.36	4.6	11.0	83.2	5.8
	1.6	1.56	1.4	6.5	86.3	7.2
190	0.1	0.064	3.5	10.3	85.1	4.6
	0.4	0.36	2.6	10.2	77.1	12.7

[a] Interpolated from the data of Baker and Krcmar (1956).
[b] Calculated values. Effective acid derived from an assumed wood ash content of 0.3% (56 meq/kg oven-dry wood), and a liquid to solid ratio of ten.
[c] 100 minus xylose in the residue and minus the xylose yield.

On a still smaller experimental scale described by Springer and Harris (1982), 0.25-mm-thick disks of southern red oak wood (sample 1 of Table I) were hydrolyzed in sealed 5-mm (outer diam.) glass ampoules heated by oil bath. Results of these hydrolyses are shown in Table III. Maximum yields of xylose in solution were calculated from the residue data and effective acid concentration as before, and are compared to experimental yields in Table IV. The calculated yields, all in excess of 80%, indicate a modest increase as the temperature is raised but a slight drop if the acid concentration is increased. The experimental values, with one exception, are somewhat greater than the calculated values, perhaps due to the protection afforded by the presence of oligomers as suggested before. The experimental values in Table IV indicate that the maximum xylose yield increases if either the temperature or acid concentration is raised. The xylose degradation (Tables II and IV) indicates that yields of furfural in these small tube studies were significant.

B. Comparison of Water and Dilute Acid Hydrolysis

The preceding data support the conclusion that maximum xylose yields increase with increased temperature and acid catalyst concentration. Analysis of data on xylose degradation (Root, 1956, 1959) leads to the conclusion that the maximum xylose yields should increase with increasing acidity if the pH is above 1.0. Since the use of water (steam) prehydrolysis is frequently proposed, it was of interest to compare the yields from water prehydrolysis, in which the pH is probably greater than 4, to that of dilute acid prehydrolysis. A comparison of the maximum xylose yields was made between water and 0.4% H_2SO_4 as hydrolyzing reagents at 170°C using aspen wood. A large difference was found--the acid solution resulted in a 79% maximum yield of xylose, the water only 61%. Other differences in the systems were also observed and the results published (Springer and Harris, 1982). It should be noted that this yield difference should decrease with increasing temperature and is perhaps quite minor at the temperatures employed in some processes (Iotech Corp., 1980).

Veeraraghaven et al. (1982) demonstrated the loss in sugar yield resulting from inadequate amounts of acid to exhaust the neutralizing capacity during hydrolysis of red oak, sorghum, and kudzu. Reduced yields which resulted from lower acidities were also observed by Conner (1983) when he studied prehydrolysis in acetic acid solutions.

TABLE III. Hydrolysis of Southern Red Oak Discs in Ampoules[a]

Temperature (°C)	H$_2$SO$_4$ added (%)	Time (min)	Residue (%)[c]	Residue components[b]				Solution components[b]		
				Glucose (%)	Xylose (%)	Mannose (%)		Glucose (%)	Xylose (%)	Mannose (%)
170	0.4	1.25	74	100	39	40		2	63	55
		2.5	68	98	21	24		2	77	69
		5.0	65	96	13	14		3	77	59
		10.0	62	95	7	20		3	78	64
		20.0	61	94	4	10		7	70	67
	0.8	1.25	67	97	24	29		2	76	75
		2.5	64	97	13	16		3	83	76
		5.0	62	94	8	22		6	79	82
		10.0	59	89	3	8		7	72	61
190	0.4	0.67	68	97	21	13		3	79	69
		1.0	63	96	12	15		3	87	81
		2.0	61	96	5	7		5	83	67
		3.0	60	91	3	11		8	82	72
	0.8	0.67	63	96	10	9		3	83	85
		1.0	60	95	6	9		5	88	75
		2.0	57	92	5	9		8	76	72

[a]Liquid to solid ratio was four.
[b]Percentages of potential available from original wood. Oligomeric forms included.
[c]Percentage of original wood.

TABLE IV. Maximum Experimental Xylose Yields Compared with Yields Calculated from Data in Table III

Temperature (°C)	H_2SO_4 Added (%)	H_2SO_4 Effective[a] (%)	Time (min)	Xylose in residues (%)[c]	Maximum xylose yield (%)	Xylose degraded[b] (%)
Experimental						
170	0.4		10.0	7	78	15
	0.8		2.5	13	83	4
190	0.4		1.5	12	87	1
	0.8		1.0	6	88	6
Calculated						
170	0.4	0.29	8.6	8	83	9
	0.8	0.69	2.8	13	81	6
190	0.4	0.29	1.7	7	85	9
	0.8	0.69	0.83	8	83	9

[a] Effective acid calculated by assuming that all the neutralizing capacity (37 meq/kg) was reacted.
[b] 100 minus xylose in the residue and minus the xylose yield.
[c] Percentage of potential available in original wood (sample 1, Table I). Oligomeric forms included.

IV. HYDROLYSIS OF OAK CHIPS IN DIRECT STEAM

Chemical principles of prehydrolysis were clarified by the ampoule experiments which were limited to isothermal conditions, constant acidity, small particle size, and fixed L/S. Digester studies made it possible to relate the data gathered in ampoules to data obtained under conditions of direct steam heating of wood chips. The major changes in going to the digester and chips were (1) the slower rate of temperature rise, (2) the changing acid concentration, both in time and space, and (3) the varying L/S. In addition, the method of acid impregnation of chips affected the acid distribution and ultimately affected the reaction rates and product yields.

Details of the hydrolyses in the digester, product yields, and techniques of analyzing for products have been reported (Scott et al., in press).

A. Chip Preparation

1. Thoroughly Impregnated Chips. An attempt was made to de-ash and thoroughly impregnate chips with acid during an extended soaking; however, the neutralizing capacity could not be completely removed even with soaking periods of a week.

In a typical preparation, 650 g of 9.5-mm southern red oak chips (Table I, sample 2, 427 g oven-dry) were submerged in 6.0 liters of 0.63% H_2SO_4, and an aspirator vacuum was drawn until gas escape was much reduced ($\sim$1 hr). The pressure was then restored to atmospheric. The chips remained submerged for 3 days, at which time the solution was decanted and replaced with 0.47% H_2SO_4. The chips were kept submerged for an additional 3 days with decantation and replacement of the 0.47% H_2SO_4 solution after 24 and 48 hours. This procedure reduced the neutralizing capacity of the chips by only 56%, from 118 to 52 meq/kg oven-dry wood. The first runs, those with numbers below 55 (Table V), were treated in this manner, but the chips for runs 76 to 81 were impregnated with 0.73% H_2SO_4. The soaking period following the last acid exchange was extended to 190 hr.

The chips used for runs 55 to 57 were first treated with 0.34% HCl using the above procedure. They were then placed in a column and washed with distilled water until free of Cl^- ($AgNO_3$ test). The washed chips were then impregnated with 0.73% H_2SO_4 as in runs 76 to 81. This dual treatment lowered the neutralizing capacity to 30 meq/kg.

For chips treated by the methods above, we found that the amount of H_2SO_4 absorbed was less than expected. The average concentration of the internal solution, calculated from the

Dilute Sulfuric Acid Prehydrolysis

sulfate and water content, was substantially lower (~10%) than that of the exterior solution. The reasons for this discrepancy is suggested to be water inaccessible to H_2SO_4, as discussed previously under the topic "xylose removal rate."

2. Rapidly Impregnated Chips. A different technique of chip preparation was used to simulate conditions reasonable for commercial impregnation of acid. In this technique chips were submerged in 2.00%, 2.25%, or 2.50% H_2SO_4. An aspirator vacuum was applied above the solution for 5 minutes after which time it was released and the chips removed from the liquid. They were then blotted, weighed, and placed in the digester to be hydrolyzed immediately. Sulfate measurements after hydrolysis showed that the acid uptake was 10 to 20% greater than would be expected from the weight of solution taken up. Since neutralization of the acid in such a short time was likely to be insignificant, the additional uptake of acid was primarily due to its diffusion into the water originally present in the green wood.

B. *Digester Operation*

Acid-impregnated woodchips were hydrolyzed in a 0.01-m^3 jacketed digester by procedures designed to simulate, as nearly as possible, conditions expected in a continuous digester. During operation the digester enclosed the removable stainless steel chip container shown in Figure 3. This was a can (A, Fig. 3) with a closed bottom; most of its wall was made of screen. The lower portion had sufficient volume to hold the liquor that drained from the chips. It was equipped with a removable false bottom (B, Fig. 3), which served to hold the chips above the liquid which accumulated in the bottom of A. Liquid could easily drain through the holes in the plate of B; the handle was a convenience for removing the residue after cooking. The lid (C, Fig. 3) was a cover to divert any water dripping from above. The chip space was 180 mm in diameter by 150 mm high and held 0.8 to 1.1 kg of impregnated chips.

The digester-to-charge mass ratio for the laboratory vessel was much greater than it would be in a commercial unit. Consequently, it was necessary to separate the two steam condensates resulting from heating digester and chips. The "condensate" consisted of liquid collected in the bottom of the digester (below container A) and liquid exhausted, during operation, from this section through a small tube. Condensates produced on the inner surfaces of the digester were reduced by preheating the interior to 100°C before admitting steam to the inside. That part of the chip condensate which

TABLE V. Average Carbohydrate Yields in Digester Runs

Run numbers	Impregnating solution H$_2$SO$_4$ (%)	Time (min)	Residue (%)[b]	Residue yields				Soluble yields		
				Glucose (%)[c]	Xylose (%)	Uronic acid[a] (%)		Glucose (%)	Xylose (%)	Uronic acid (%)

Thoroughly Impregnated Chips – Southern Red Oak

79	0.47	4	69	99	18	26		2	82	75
43,52,80	0.47	6	66	96	13	16		3	84	76
40,44,53,81	0.47	9	64	93	10	11		4	84	65
41,45,48,49	0.47	12	65	95	7	–		5	81	52
37,42,54	0.47	15	64	95	7	6		5	82	30
38,50	0.47	18	64	96	6	–		6	82	24
39,51	0.47	21	65	96	6	5		6	80	18
76	0.73	4	66	96	11	16		3	86	80
55,77	0.73	6	65	95	10	12		5	86	73
56,78	0.73	9	63	93	6	6		6	81	54
57	0.73	12	62	93	5	5		6	80	43

Thoroughly Impregnated Chips - Aspen						
68	0.73	6	66	95	7	10	4	82	80

Thoroughly Impregnated Chips - Aspen								
68	0.73	6	66	95	7	10	4	82	80
Thoroughly Impregnated Chips - Birch								
69	0.73	6	64	97	11	13	3	86	84
Rapidly Impregnated Chips - Southern Red Oak								
70	2.0	6	67	94	12	14	5	79	72
71	2.0	9	65	95	9	12	6	78	56
72	2.0	12	65	94	11	9	7	76	45
73	2.25	6	67	96	8	12	6	80	68
74	2.25	9	66	96	6	10	7	75	52
75	2.25	12	65	94	8	9	7	72	42
63	2.5	6	64	96	10	11	6	81	67
64	2.5	9	62	92	5	7	7	75	48
65	2.5	12	62	91	7	6	9	71	37

[a] Residue uronic acid averages do not include runs 37 through 50.
[b] Percentage of oven-dry wood charged.
[c] Percentage of potential available in wood charged. Oligomeric forms included.

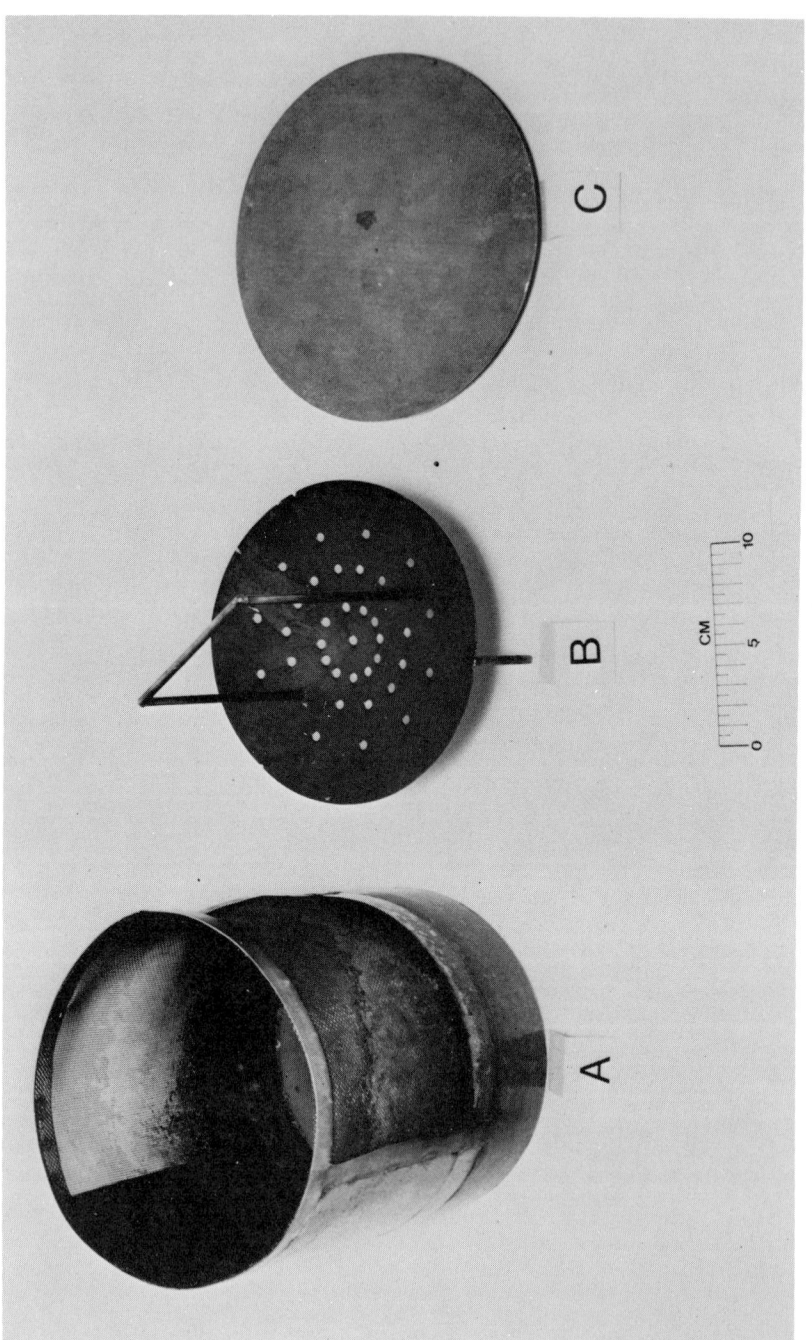

FIGURE 3. Chip container assembly for hydrolysis of wood chips; A, container; B, insert to support chips; C, cover.

was not retained by the chips, but flowed over them, was collected in the bottom of container A as "free liquor." The term "extract" will refer to the liquid retained by the hydrolyzed chips (the "residue"), and subsequently removed with the additional water necessary to elute solutes from the chips.

C. Water Movement

Comparison of the results of direct steam heating with those obtained in ampoule studies required that hydrolyses be at the same temperature, time, and catalytic acid concentration. The latter was very difficult to establish because it depended on many factors. Among these was the movement and location of water in the digester and within the chip bed. Initially, the effect of the condensing steam was studied by minimizing the influence of other variables. Hydrolysis of 9.5-mm chips with an estimated heatup time of less than a minute minimized the effect of varying temperature. A long equilibration with acid reduced the neutralizing effect of ash.

The calculations of water movement in the digester, shown in Table VI, include measured and calculated values. Measured values are: water in the charge (col. 3), water in the cooked chips after discharge (col. 7), and water in the free liquor after discharge (col. 8). Knowing the heat capacity of the charge and the temperature rise and assuming that heating occurred by the condensation of saturated steam, the total water at cooking temperature (col. 6) was calculated. Assuming that cooling on discharge occurred by adiabatic vaporization, the total water at discharge (col. 9) was obtained. Similarly, the water content, at reaction temperature, of the chips and the free liquor (cols. 4 and 5) was calculated from the measurements of their mass and water content after discharge. The values are approximations, as can be seen by comparing the sums of water in the chips and in free liquor to the total water. However, it can be concluded that the free liquor accumulated very early in the cooks. Notice that there was little if any increase in the amount of free liquor as the time of the cook was extended. This means that there was very little moisture dripping from the chip bed after the initial heatup interval, that is, after the onset of hydrolysis. This is substantiated by the data in Table VII which indicate that only 4 to 9% of the solubilized solids appeared in the free liquor. Runs 60 to 62 (Table VI) were designed to determine how rapidly the free liquor accumulated, and they do indeed show that at the end of the heatup time, estimated to be 1 min, the accumulation of free liquor was essentially complete.

The data of Table VI also show the suprising result that heating the chips led to a reduction in the amount of water

TABLE VI. Average Water Distribution in Digester Runs[a]

Number of runs (1)[b]	Time (min) (2)	Water in charge (%) (3)[c]	Water during hydrolysis (%)			Water after discharge (%)		
			In chips (4)	Free liquor (5)	Calculated total (6)	In chips (7)[c]	Free liquor (8)[c]	Calculated total (9)

Thoroughly Impregnated Chips - Southern Red Oak

2	4	132	122	63	183	89	50	137
5	6	128	109	81	179	80	64	134
6	9	126	103	74	176	74	58	132
5	12	122	97	71	170	70	57	128
3	15	125	96	79	174	69	63	131
2	18	127	100	79	177	74	55	134
2	21	120	101	83	168	73	66	127

Thoroughly Impregnated Chips – Aspen								
1	6	184	177	105	251	132	83	191
Thoroughly Impregnated Chips – Birch								
2	6	169	162	98	231	121	77	176
Rapidly Impregnated Chips – Southern Red Oak								
1 (run 60)	1	110	103	68	154	74	96	115
1 (run 61)	2	113	99	66	158	71	52	117
1 (run 62)	4	113	94	72	159	66	57	117
3	6	113	96	66	158	69	52	118
3	9	114	93	69	159	67	55	119
3	12	113	90	70	159	64	55	117

[a] All digester runs were at 170°C. Values are percentages of oven-dry wood charged.

[b] Data from runs differing in acid concentration were combined because the differences did not influence water distribution. Run numbers are listed in Table V.

[c] Measured values; others were calculated.

TABLE VII. Dissolved Solids, Acid Movement, and Acid Neutralization in Digester Runs[a]

Run numbers	Impregnating solution H_2SO_4 (%)	Time (min)	Dissolved solids		Sulfate[b]		Neutralizing capacity			Calculated effective [H^+]	
			Total (kg)	In free liquor (%)[c]	Charged (mmol)	In free liquor (%)[c]	Charge (meq)	Residue (meq)	Calculated (pH)	Equivalent H_2SO_4 (%)

Thoroughly Impregnated Chips – Southern Red Oak											
79	0.47	4	0.31	6	56	29	37	22	1.60	0.25
80	0.47	6	0.33	7	56	29	37	25	1.55	0.28
81	0.47	9	0.35	8	55	30	37	21	1.63	0.23
55	0.73	6	0.36	4	81	30	30	18	1.37	0.42
56	0.73	9	0.37	6	81	33	30	15	1.43	0.36
57	0.73	12	0.38	6	81	32	30	10	1.46	0.34
76	0.73	4	0.34	6	88	28	38	29	1.30	0.49
77	0.73	6	0.34	9	89	30	38	25	1.35	0.44
78	0.73	9	0.36	9	89	30	38	19	1.36	0.43

Rapidly Impregnated Chips - Southern Red Oak

70	2.0	6	0.33	7	144	39	118	27	1.41	0.38
71	2.0	9	0.35	9	147	40	118	23	1.46	0.34
72	2.0	12	0.35	7	138	43	118	20	1.61	0.24
73	2.25	6	0.33	9	175	38	118	28	1.23	0.58
74	2.25	9	0.34	8	180	38	118	21	1.21	0.60
75	2.25	12	0.35	8	156	41	118	24	1.38	0.41
63	2.5	6	0.36	5	176	40	118	24	1.25	0.56
64	2.5	9	0.38	4	180	41	118	17	1.25	0.56
65	2.5	12	0.38	4	174	46	118	16	1.34	0.46

[a] Basis: 1 kg of oven-dry wood charged.

[b] Total of sulfate and bisulfate, determined by barium precipitation.

[c] Percentage of the column to the left.

retained by the chips rather than the expected increase, (compare cols. 3 and 4). Table VII indicates that 30 or 40% of the $SO_4^=$ was transferred to the free liquor depending on the method of impregnation. As discussed above, this must have occurred during the heatup period. Consideration of these facts led to the conclusion that liquid was forced from the interior of the wood and carried by condensate to the free liquor. A plausible explanation for the driving force which moved this liquid from within the chips is that residual air was present. Water in the interior of the chip exerted a partial pressure equal to its vapor pressure, which, after heatup, was the same pressure as that at the exterior. The presence of air increased the total pressure in the interior since its partial pressure was added to that of the water. Thus, to establish pressure equilibrium, the air must have been expelled from the chips, and in its movement to the surface forced out much of the liquid.

This movement of solution from the interior of the chip to the surface, if properly interpreted, must depend on many factors. These would include the method of impregnation, the wood morphology and the possibility of the liberation of gases during heatup. Tables VI and VII indicate quite large effects due to differences in the method of impregnation. After heatup the rapidly impregnated chips retain less water and a substantially lower portion of the charged sulfate. Both aspen and birch showed a loss of water after heating but the effect was not nearly as great as that for oak (Table VI).

D. Effective Acid Strength

Lacking a direct measurement, the acidity of the hydrolyzing solution can be estimated from its average composition, specifically the content of water, cations and sulfate. These quantities are obtained from a knowledge of the total quantities charged to the digester and the distribution of each between the various phases after reaction. Measurements of the distribution of cations and sulfate were only made for later runs; the data are reported in Table VII. The procedure for calculating the effective acidity from this information is demonstrated using the following calculation for run 80. The quantities are based on 100 gms of wood charged.
Components in extract:
    Water (Table VI, col. 4): 109 g or 0.109 L at 20°C
    Cations released into extract (Table VII):
        = 3.7-2.5 = 1.2 meq or 11.0 meq/L.

Bisulfate ion (Table VII): Assumes no secondary ionization, all sulfate is present as bisulfate.

= millimols HSO_4^- in charge - millimols HSO_4^- in free liquor

= 5.6-(5.6 × 0.29) = 3.98 meq or 36.5 meq/L

From a charge balance,

[H^+] + [cations] = [HSO_4^-]

[H^+] = 36.5-11.0 = 25.5 meq/L or 0.0255 eq/L

pH = $-\log_{10}[0.0255]$ = 1.59

The value in Table VII differs from this value because the table calculations included the ionization constant for the bisulfate ion. This is thought to be an unnecessary refinement even at this relatively high pH.

It will be noted that all of the cations released from the residue were assumed to be present in the extract, that is, none were transferred to the free liquor. Three measurements of cations in the free liquor from thoroughly impregnated chips showed so little that, for calculation, they were assumed to be zero. For the case of the rapidly impregnated chips, the few measurements indicated that, on the average, 15% of the cations released were transferred to the free liquor.

Table VII shows that there is a great difference between the concentration of acid used for impregnation and the acidity of the hydrolyzing solution. There is almost a 50% reduction in the acid strength for thoroughly impregnated chips. For the case of the rapidly impregnated chips, it is apparent that little meaning can be attached to the strength of the impregnating solution without considering the other factors that influence the final solution strength. The movement of water and other components in the charge would have important consequences in the design of process equipment.

It should also be noted, from Table VII, that the neutralizing capacity of the residue tends to decrease as the interval of reaction increases but it is never completely consumed.

E. Product Yields

1. *Xylose from Thoroughly Impregnated Chips.* The two data groups for the two acid levels, 0.47% and 0.73% H_2SO_4, were similarly correlated and examined in the following manner, which is described in detail for the first group.

Correlation of the xylose removal data, for the 0.47% H_2SO_4 group, resulted in the empirical expression:

XR = xylose remaining in the residue, % of original
   = 86.5 exp(-0.863t) + 13.5 exp(-0.044t)

Here t is the time interval after the onset of hydrolysis. Because the heatup time for the digester charge is 1 minute, t is 1 minute less than the measured time interval in the digester. The lower curve of Figure 4, which includes the original data from Table V, is a graphical presentation of the equation. As previously emphasized, the equation is empirical and its use is an expediency to allow mathematical manipulation of the data.

Using a value of 0.19% H_2SO_4 for the effective acidity, and the above expression XR for the release of xylose to the solution, a numerical integration (sec. II, B) led to the yield of xylose in solution represented by the upper curve of Figure 4. The maximum point on this curve is 85.2% at 7.2 minutes. The maximum experimental value was 84 at 6 and 9 minutes (Table V).

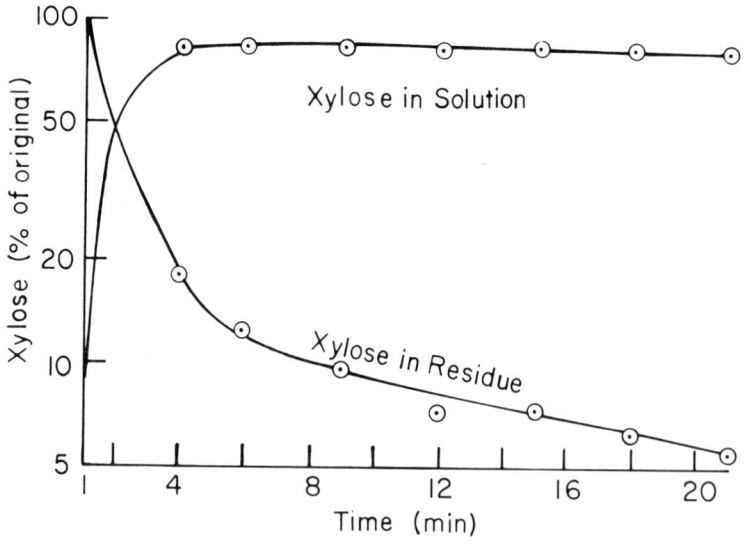

FIGURE 4. *Hydrolytic separation of original wood xylose into soluble and residue fractions. Conditions: southern red oak chips thoroughly impregnated with 0.47% H_2SO_4, 170°C. Data points from Table V.*

Dilute Sulfuric Acid Prehydrolysis

The acidity value used for this data group, a total of 19 runs at 7 different times, is somewhat less than that given in Table VII for runs 79, 80 and 81. The wood charge for the earlier runs had a neutralizing capacity of 52 meq/kg rather than the 37 meq/kg indicated for runs 79, 80, and 81. Accounting for this lowered the acidity appreciably, resulting in an average effective acidity for the entire group of 0.19% H_2SO_4.

The experimental yields of solubilized xylose shown in Figure 4 appear to be very near the calculated curve. However, the logarithmic plot obscures the fit; a much more stringent test of the data is made in Figure 5. Here the ordinate values, percentage xylose degraded, are obtained by subtracting the sum of recoveries in solution and residue from the original 100%. This value is very sensitive to experimental error for it is the difference between two large numbers; 1% error in determining the solution composition results in 15 to 20% error in the amount of xylose degraded. However, there is little doubt that the measured xylose losses are close to those predicted by calculation, which demonstrates that the assumed model satisfactorily correlates the experimental data.

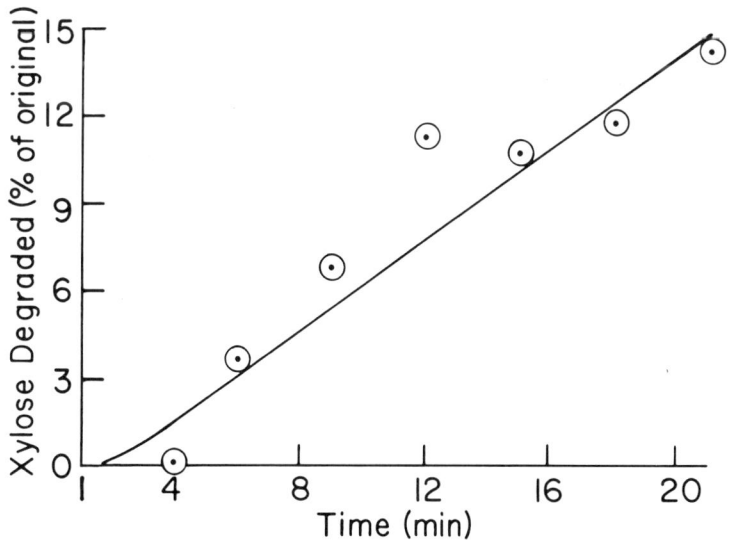

FIGURE 5. *Loss of original wood xylose to other products. Curve and points calculated from Fig. 4.*

The data group for the runs in which the material was impregnated with 0.73% H_2SO_4 was treated in the same manner, with the results shown in Figures 6 and 7. In this case the effective acidity was taken as the average of the values listed in Table VII, 0.41% H_2SO_4, and the removal curve was represented by the function:

$$XR = 89.3\exp(-1.080t) + 10.7\exp(-0.0656t)$$

The calculated maximum solubilized xylose yield was 85.5% occurring at 5.2 minutes. This may be compared with the experimental maximum of 85.7% at 6 minutes. In this case, as with the prior group, the calculated yields are very close to those measured. It was expected, because of the presence of oligomers, that the experimental values would exceed those calculated. However, other factors such as the distribution of acid throughout the chip could have an opposing effect on the actual yield.

One may compare the data from direct steam heating to that from ampoule experiments. As shown in Table III, at 170°C the maximum xylose yields obtained in ampoules using 0.4% and 0.8% H_2SO_4 (0.29% and 0.69% effective acid) were 78 and 83% respectively. The experimental maximums obtained in the digester were 83.8% with 0.19% effective H_2SO_4 and 85.7% with 0.41%. Both pairs indicate an increase in maximum xylose yield resulting from an increase in acidity. The ampoule and digester studies used different wood samples, employed different analytical procedures, and were performed by different personnel. The reported higher yields from the digester are probably not significant and it was concluded that xylose yields were not reduced by direct steaming per se.

2. *Xylose from Rapidly Impregnated Chips.* Analysis of the data for the rapidly impregnated chips by the procedure used for previous runs is quite unsatisfactory. The calculated acid concentrations, reported in Table VII, are much too low to account for the rapid degradation of xylose. Measurements of furfural formation, Table VIII, substantiate the high xylose losses. Evidently the acid is not uniformly distributed throughout the substrate. This would be expected considering the short impregnation and equilibration times. The unequal distribution of acid has the following consequences, which are supported by the data:

(a) The quantity of xylose removed from the residue (in the range of 90% removal) is less when the acid is not uniformly distributed. That part of the xylan not in contact with acid will hydrolyze at a reduced rate, and much of this material will remain in the residue.

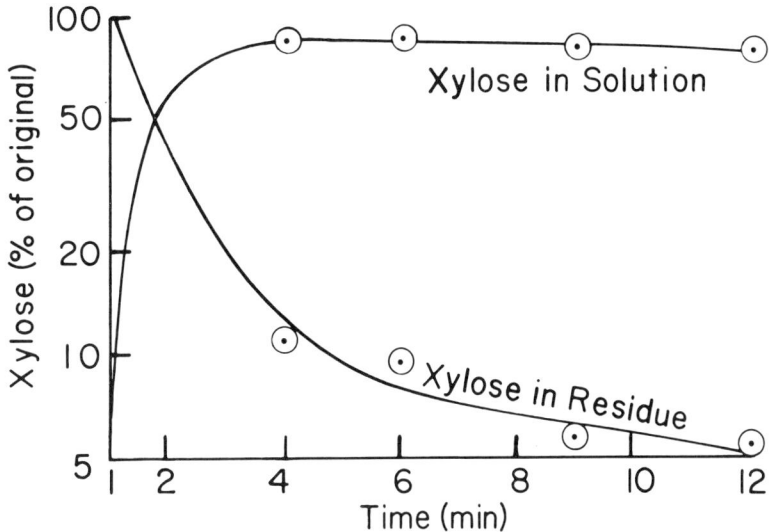

FIGURE 6. Hydrolytic separation of original wood xylose into soluble and residue fractions. Conditions: southern red oak chips thoroughly impregnated with 0.73% H_2SO_4, 170°C. Data points from Table V.

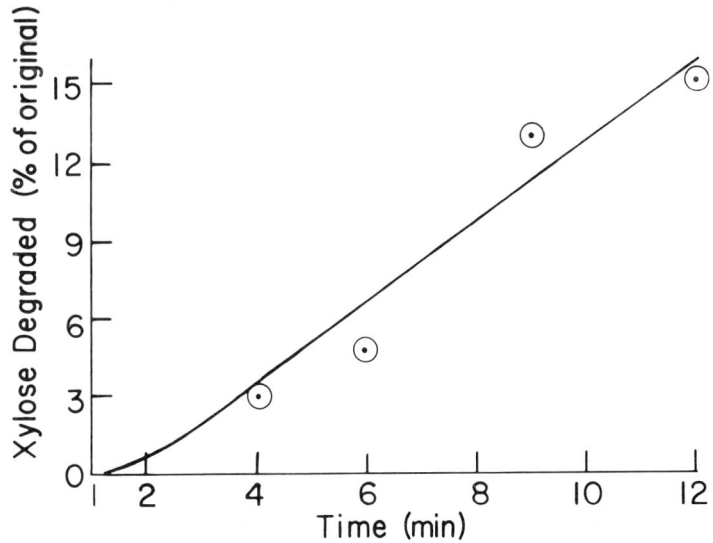

FIGURE 7. Loss of original wood xylose to other products. Curve and data points calculated from those of Fig. 6.

TABLE VIII. Furfural Yields in Digester Runs

Run numbers	Impregnating solution H_2SO_4 (%) (1)	Time (min) (2)	Xylose reacted (%) (3)[a]	Uronic acid reacted (%) (4)[a]	Potential furfural (%)		Furfural yield (%)	
					From xylose (5)[b]	From uronic acid (6)[b]	Potential (7)[b]	Actual (8)[b]

Thoroughly Impregnated Chips

79	0.47	4	0.0	0.0	0.0	0.0	0.0	0.0
52,80[c]	0.47	6	3.6	7.3	0.49	0.13	0.62	0.46
53,81[c]	0.47	9	5.1	23.4	0.69	0.42	1.11	0.80
51	0.47	21	14.1	71.2	1.89	1.26	3.15	9.93
76	0.73	4	3.2	4.3	0.43	0.08	0.51	0.55
55,77[c]	0.73	6	4.7	16.0	0.63	0.29	0.92	0.84
56,78[c]	0.73	9	13.0	38.3	1.76	0.69	2.45	1.32
57	0.73	12	15.1	51.3	2.04	0.92	2.96	1.57

Rapidly Impregnated Chips

70	2.0	6	8.9	14.0	1.19	0.25	1.44	1.04
71	2.0	9	12.8	32.0	1.71	0.57	2.28	1.43
72	2.0	12	12.9	46.4	1.73	0.82	2.55	1.87
73	2.25	6	11.7	20.4	1.57	0.36	1.93	1.16
74	2.25	9	19.0	38.5	2.54	0.68	3.22	1.93
75	2.25	12	20.0	49.6	2.68	0.88	3.56	2.18
63	2.5	6	9.3	21.8	1.24	0.39	1.63	1.37
64	2.5	9	19.8	44.7	2.65	0.79	3.44	1.97
65	2.5	12	22.4	56.7	3.00	1.01	4.01	2.46

[a]Percentage of potential available in wood charged.
[b]Percentage of oven-dry wood charged.
[c]Data are averages of the two runs.

b. The xylose loss is much greater. If the acid is poorly distributed, the xylose solubilized is released into areas of higher than average acidity. Although there is somewhat less xylose brought into solution (see a), the higher acid concentration results in substantially greater losses than found in a system with a uniform distribution.

c. The maximum xylose yield is decreased. This results from the slower release and more rapid decomposition of the xylose, and is evident from the data of Table V where a decrease of approximately 5% is indicated.

d. Furfural production is greater. This results from the increased xylose decomposition. The maximum combined yield of xylose and furfural from the nonuniformly distributed acid medium is only 1 to 2% less than that from the system in which the acid is uniformly distributed.

The distribution of acid throughout the reacting substrate is undoubtedly one of the important factors determining the rate of xylose removal and, consequently, xylose and furfural yields. The wood species, chip size, and method of impregnation, all influence the acid distribution. Detrimental effects due to nonuniformly distributed acid are apparent in the data gathered using 915-mm chips. Even greater effects would be expected when using larger chips.

3. Furfural and Acetic Acid Yields. Since furfural is a degradation product of both xylose and uronic acid, it is unavoidably produced in the prehydrolysis. Experimental yields are shown in column 8 of Table VIII. The extent of the reaction of both xylose and uronic acid is also shown; these values are obtained by summing the recoveries in the extract and residue (Table V), and subtracting from the charge. The relative importance of each as furfural sources, if they were converted in stoichiometric yield, is indicated by the potential furfural yields in columns 5 and 6 of Table VIII. It is apparent that uronic acid could contribute appreciably to the furfural yield. However, uronic acids degrade to products other than furfural (Feather and Harris, 1973) and yields probably do not exceed 30% of theoretical. Unfortunately, information on furfural from uronic acid is unavailable, making it difficult to interpret the furfural yields, but projected yields from the reacting xylose can be calculated and compared with the experimental values (Table VIII, cols. 5 and 8).

It was pointed out that initial furfural formation is stoichiometric (based on xylose reacted) and decreases as the reaction proceeds. This is shown in Figure 8 where curves were calculated at the approximate conditions of the prehydrolysis (L/S = 1.1, pH = 1.45, 170°C) using the procedure described in section II-C. The integrated value is 100 × furfural

Dilute Sulfuric Acid Prehydrolysis

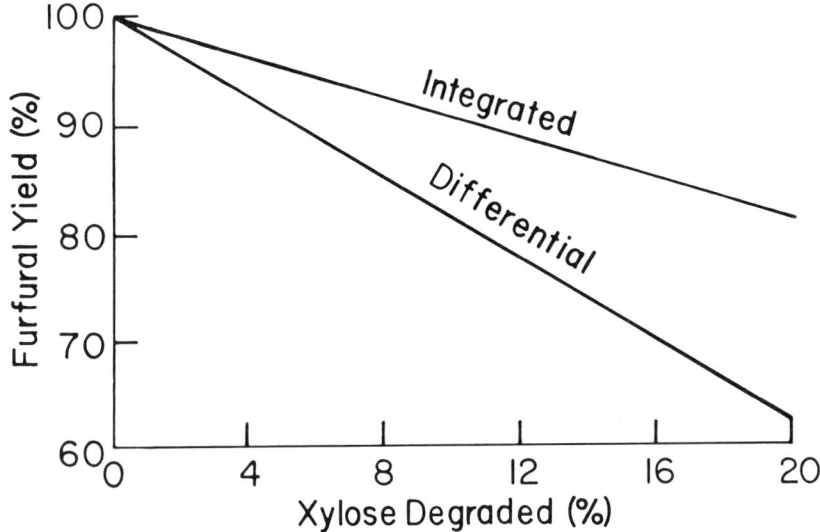

FIGURE 8. Calculated furfural yields, as mol percentage of reacted xylose.

produced/xylose reacted over the entire reaction period, while the differential value is the same ratio but based on an infinitesimal interval. The differential percentage assumes negative values after the point of maximum furfural yield (based on charge) is reached.

The integrated yield values in Figure 8 were used to calculate the curve in Figure 9 where the yield is based on wood charged. The experimental points in Figure 9 are scattered about the predicted yield based on the xylose alone. For two reasons it would be expected that they would be above the predicted curve: (1) uronic acid and arabinose must make contributions to furfural yield and (2) much of the furfural escapes from the extract and free liquor to the condensate where the degradation rate would be less. In general, more than half the furfural is recovered in the condensate. Consideration of the data in Figure 9 leads to the conclusion that the yield of furfural can be estimated from the calculated curve.

The acetyl content of the wood used in the digester cooks was 4.30% (Table I, sample 2), which is a potential acetic acid yield of 6.00% on the wood. It had been assumed that the acetyl should either be carried into solution attached to the xylan or be cleaved in situ and solubilized as acetic acid, in which case, the ratio of acetyl to xylose in the residues

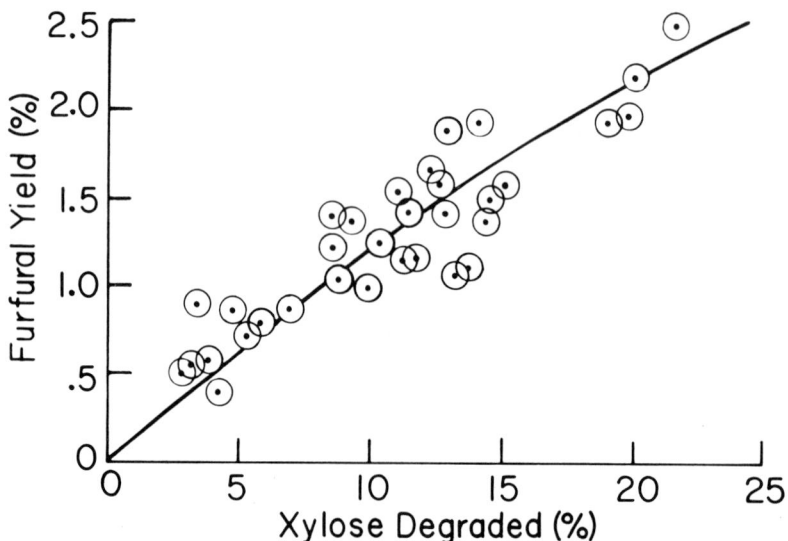

FIGURE 9. Furfural yields as percentage of wood. Integrated yield values of Fig. 8 used to calculate curve.

would be less than or equal to that ratio in the original wood. This supposition seems to be erroneous. Measurements of the distribution of acetyl following hydrolysis in runs after 69 disclosed that only 75 to 85% of the wood acetyl was in the liquid phases. A measured 11 to 20% was still in the residue and thus at a higher mole ratio (average 1.0) to xylose than in the original wood (ratio, 0.72).

VII. SUMMARY

This report of hemicellulose removal from southern red oak provides new information on wood hydrolysis.[1] Small scale studies with finely divided wood produced kinetic data on xylose removal and degradation. Direct steam heating of acidified wood chips in a small digester resulted in data more applicable to commercial operations. We summarize our observations in the following conclusions:

[1] A more extensive analysis and further computations are included in a report by Harris et al. (1983).

(1) High yields of soluble xylose from hydrolysis at 170°C are obtained only at a low pH requiring a strong acid. Hydrolysis without the addition of such an acid results in greatly reduced yields.

(2) Xylose removal rates together with xylose decomposition rates permit one to predict xylose and furfural yields.

(3) Maximum xylose yields increase if either the temperature or the acid concentration is raised. These yields (including xylose in oligomers) ranged up to 86% of the potential xylose in oak chips.

(4) Chip-size wood causes difficulties in controlling and estimating the important variables: temperature, liquid-to-solid ratio, and pH. The latter is least controllable. Green chips absorb relatively more acid than water during a brief 5-minute vacuum impregnation. However, acid solution is subsequently expelled from the chips during initial heating with direct steam. As a consequence, 40% of the impregnated acid may not be available in the wood when hydrolysis begins.

(5) Carboxylate groups in the raw material combine with hydrogen ions of impregnated acid before and during hydrolysis. This loss of acidity can be appreciable at the desired conditions of dilute acid and low amounts of solution. The amount of acid lost in this manner can be measured by titrating the alkalinities of ash from the raw material and its residue.

(6) The effective pH was estimated from acid neutralized, acid expelled, and calculated water movement during steam heating. The estimated pH approximately agreed with that expected from measured xylose and furfural yields.

(7) Product yields on the basis of the dry weight of oak chips were: 15 to 17% xylose, 1.0 to 2.2% furfural, and 4.5 to 5.1% acetic acid. About 20% of the xylose yield was in oligomers. The conditions at which these yields were obtained were 170°C for 6, 9, or 12 minutes with chips which had been vacuum impregnated with 2.0 or 2.25% H_2SO_4 for 5 minutes.

(8) In the early stages of hydrolysis, furfural yields can be increased without a decrease in soluble xylose because the latter is being replenished from the wood.

(9) Long equilibration of wood with acid solution leads to a lower than expected uptake of acid. This is presumed due to inaccessible water in the wood. It may be that xylan which survives long hydrolysis is also inaccessible to acid.

REFERENCES

Baker, A. J., and Krcmar, G. F. 1956. "Kinetics of the Prehydrolysis of Wood." Unpublished Report. U.S. Forest Service, Forest Products Laboratory, Madison, Wisconsin.

Conner, A. H. 1983. "Prehydrolysis of Southern Red Oak With Water and Acetic Acid Solutions." In press.

Feather, M. S., and Harris, J. F. 1973. Dehydration Reactions of Carbohydrates. *Adv. Carbohydr. Chem. Biochem.* 28: 161-224.

Harned, H. S., and Owen, B. B. 1958. "The Physical Chemistry of Electrolytic Solutions." Reinhold Pub. Corp., New York.

Harris, J. F., Conner, A. H., Jeffries, T., Minor, J. L., Pettersen, R. C., Scott, R. W., Springer, E. L., and Wegner, T. H. 1983. "Two Stage, Dilute Sulfuric Acid Hydrolysis of Wood." FPL Research Paper. U.S. Forest Service, Forest Products Laboratory, Madison, Wisconsin.

Harris, J. F., Saeman, J. F., and Locke, E. G. 1963. In "The Chemistry of Wood" (B. L. Browning, ed.), Chap. 11. Interscience, New York.

Iotech Corporation. 1980. "Optimization of Steam Explosion Pretreatment." Final Report, DOE Contract No. DE-AC02-79ETZ, 3050.

Krasowski, J. A., and Marton, J. 1982. The Formation of Oxalic Acid During Bleaching--A Source of Deposits. Tappi Proceedings. Research and Development Division Conference, Ashville, N.C.

Lee, Y. Y., and McCaskey, T. A. 1982. Hemicellulose Hydrolysis and Fermentation of Resulting Pentoses to Ethanol. Tappi Proceedings. Research and Development Division Conference, Ashville, N.C.

Root, D. F. 1956. "Kinetics of the Acid Catalyzed Conversion of Xylose to Furfural." PhD. Thesis (Chemical Engineering), University of Wisconsin, Madison.

Root, D. F., Saeman, J. F., Harris, J. F., and Neill, W. K. 1959. Kinetics of the Acid-Catalyzed Conversion of Xylose to Furfural. *Forest Products Journal* 9(5):158-164.

Scott, R. W., Wegner, T. H., and Harris, J. F. 1983. "Dilute Sulfuric Acid Prehydrolysis of Southern Red Oak Chips Using Direct Steam Heating." In press.

Sookne, A. M., and Harris, M. 1940. The Relation of Cation Exchange to the Acidic Properties of Cotton. *Textile Research Journal* 10:405-419.

Springer, E. L. 1966. Hydrolysis of Aspenwood Xylan With Aqueous Solutions of Hydrochloric Acid. *Tappi* 49(3):102-106.

Springer, E. L. Unpublished.

Springer, E. L., and Harris, J. F. 1982. Prehydrolysis of Aspen Wood With Water and With Dilute Aqueous Sulfuric Acid. *Svensk. Papperstidning* 85(15),R152-R154.

Springer, E. L., and Zoch, L. L. 1968. Hydrolysis of Xylan in Different Species of Hardwoods. *Tappi* 51(5):214-218.

Tout, R. E., Gilboy, W. B., and Spyrou, M. M. 1976. Neutron Activation Studies of Trace Elements in Tree Rings. *J. Radioanal. Chem.* 37:705-715.

Veeraraghavan, S., Chambers, R. P., Myles, M., and Lee, Y. Y. 1982. "Kinetic Modeling and Reactor Development for Hemicellulose Hydrolysis." AICHE National Meeting, Orlando, Fla.

THE ENERGY COSTS OF INCREASED
ORGANICS RECOVERY FOR CHEMICAL BY-PRODUCTS
IN KRAFT PULP MILLS

W. J. Frederick, Jr.

Department of Forest Products
Oregon State University
Corvallis, Oregon

I.	INTRODUCTION	143
II.	CHEMISTRY AND ENERGY BALANCES	144
III.	COGENERATION	149
IV.	PULP MILL OPERATING PRACTICES AND ENERGY CONVERSION	152
V.	COST ANALYSES OF ORGANICS RECOVERY	153
VI.	SUMMARY AND CONCLUSIONS	159

I. INTRODUCTION

In chemical pulping of wood, about half of the raw material is converted to a useful papermaking fiber. The remainder of the wood is dissolved in a solution of the pulping chemicals. An important aspect in the operation of a modern chemical pulp mill is the recovery of not only the pulping chemicals but also the chemical energy from dissolved wood organics in the spent pulping liquor. In all commercial chemical pulping processes, the energy is recovered by burning concentrated spent liquor in boilers to produce steam. The steam produced is used for heating in the pulping and papermaking processes, and in most instances, for generating electricity (cogeneration). Today, energy recovered from spent pulping liquors accounts for 37.4% of all energy consumed by the pulp and paper industry (Grant and Slinn, 1983).

Some of the dissolved wood organics are recovered from spent pulping liquors before the liquor is burned (Casey, 1980). The practice is more common in sulfite pulp mills, where vanillin, ethanol, Torula yeast, and lignosulfonates are either recovered directly from the spent liquor or are produced from chemicals extracted from it. The main by-products from kraft spent liquor are turpentine and tall oil. Although alkali lignin is recovered from kraft spent liquor, the amount is small relative to the total quantity available. There has been considerable interest in using alkali lignin for adhesive extenders or as a chemical feedstock for phenol production. If either of these markets are developed, they could become large enough to change significantly the use of dissolved wood organics as fuel.

Part of the costs of recovering dissolved wood organics from spent liquors is the cost of replacing them with another fuel. The energy tradeoffs are not always obvious because of the complex relationship between composition of the spent liquor burned in the recovery boiler, its steam generating capacity, steam requirements for process heating, and electrical energy generation. This report discusses in general terms how changes in any part of the pulping process can affect a mill's energy balances and purchased fuel requirements, and specifically how to determine the cost of replacement energy that is incurred when a fraction of the dissolved wood organics is recovered for by-products instead of used as fuel.

II. CHEMISTRY AND ENERGY BALANCES

Chemical pulping results in the selective removal of lignin relative to carbohydrates from lignocellulosic materials. In the kraft pulping process, the active pulping chemicals are sodium hydroxide (NaOH) and sodium hydrosulfide (NaHS). NaOH is responsible for the hydrolysis and dissolution of the lignin macromolecule. The role of NaHS is to protect the carbohydrates from rapid degradation by alkaline hydrolysis (Rydholm, 1965).

An important aspect of kraft pulping technology is the recovery of the active pulping chemicals. The kraft pulp mill consists of a closed sodium-sulfur cycle and a closed calcium cycle (Figure 1). The chemicals of the sodium-sulfur cycle leave the digester as components of the spent (black) liquor. They are recovered as sodium carbonate (Na_2CO_3), sodium sulfide (Na_2S), and sodium sulfate (Na_2SO_4) after combustion of the black liquor. When they are dissolved in water, Na_2S

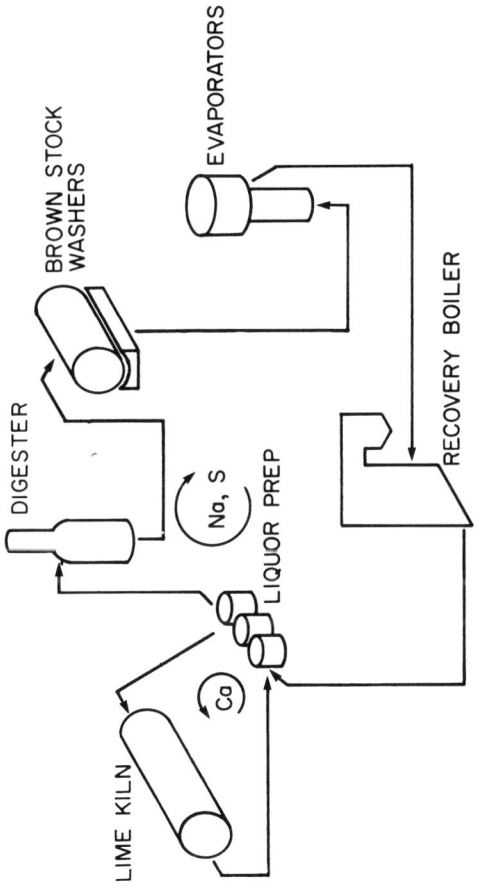

FIGURE 1. Chemical cycles in kraft pulp mill (Na = sodium, S = sulfur, Ca = calcium).

hydrolyzes to yield NaOH and NaHS. The solution is then causticized with calcium hydroxide ($Ca(OH)_2$) to convert Na_2CO_3 to NaOH. The resulting calcium carbonate is recycled through a calciner and the product calcium oxide is slaked to form $Ca(OH)_2$. Some of the NaHS is oxidized to sodium thiosulfate ($Na_2S_2O_3$) during the causticizing process and during pulping. Because of thermodynamic limitations, conversion inefficiencies, and accumulation of inerts, the pulping liquor entering the digester contains quantities of Na_2CO_3, Na_2SO_4, $Na_2S_2O_3$, and potassium and chloride salts, as well as the active pulping chemicals. The ratio of total inorganic chemicals to active pulping chemicals plays an important role in determining the energy recovery efficiency in kraft pulp mills (Frederick, 1981).

The recovery boiler is the heart of the energy conversion system in a kraft pulp mill. It functions as both a chemical reactor and steam generator, accomplishing: (1) separation of organics from inorganics, (2) conversion of the organics to CO_2 and water vapor, (3) reduction of sodium-sulfur compounds to Na_2S, and (4) recovery of energy released during combustion.

There are two important limitations related to energy recovery in the operation of a recovery boiler. The first is that the liquor fired in the boiler must be of high enough quality to sustain autogeneous combustion. This requires that the liquor be concentrated to at least 60% solids content and the heat of combustion of the dissolved solids in the liquor exceed a minimum value. Experience has shown that the heat of combustion of the dissolved solids must not fall below about 12,600 kJ/kg or auxiliary fuel is required to sustain combustion. This assumes normal operating conditions of 60-65% solids content at the boiler's rated thermal load. The second limitation is related to boiler overheating. Again, experience indicates that recovery boilers cannot be fired at more than 15-20% above their designed thermal load. The consequences of exceeding this limit are rapid plugging of the boiler and/or severe corrosion (Tran et al, 1983; Reeve et al, 1983). Note that these limits should be taken as guidelines only; the actual limit may differ from boiler to boiler.

The effectiveness of the recovery boiler as a steam generator is best expressed as its thermal efficiency, which is defined as the ratio of heat recovered as steam to the total energy as fuel and sensible heat entering the recovery boiler. Thermal efficiency is calculated from mass and energy balances on the recovery boiler. The TAPPI Recovery Boiler Test Procedure Short Fort (TAPPI, 1979) provides a convenient method for the calculations. Table I shows the heat balance for a typical recovery boiler. The thermal efficiency for most recovery boilers is between 60 and 70%.

TABLE I. Heat Balance and Thermal Efficiency For a Kraft Recovery Boiler

Heat inputs (kJ/kg dry solids)	
Heat of combustion of dry solids	15,290
Sensible heat of liquor	290
Sensible heat of combustion air	930
Total heat input	16,510
Heat losses (% of input)	
Sensible heat of dry flue gas	5.92
Sensible heat of moisture in air	0.14
Heat losses as water vapor	
Hydrogen in liquor solids	6.90
Water in black liquor	8.98
Heat loss as molten salts	3.45
Loss from reduction of sulfur salts	
Makeup	0.00
Recycled	5.32
Radiation loss	0.30
Unaccounted losses	2.50
Total heat loss	33.51
Thermal efficiency, %	66.49

There are three general components of kraft black liquor: dissolved wood organics (40-50% by weight), inorganic pulping chemicals (20-30% by weight), and water (30-40% by weight). These components determine the heating value of black liquor. Wood organics are the source for essentially all of its heat of combustion. They are the only component in black liquor that are truly combustible. The other components act as inert diluents with respect to its heat of combustion. They lower the overall fuel quality by decreasing the average heat of combustion and absorbing 10-15% of the heat of combustion of the organics as latent and sensible heat (Table I). When one considers the fuel characteristics of black liquor in this way, the impact of pulp mill operation on energy recovery becomes evident very quickly.

The dissolved wood organics are comprised of specific groups of compounds that each contribute their own heats of combustion: carbohydrates (cellulose and hemicelluloses) (17,200 kJ/kg), softwood lignin (26,400 kJ/kg), hardwood lignin (24,700 kJ/kg), extractives (39,300 kJ/kg) (Annergren et al, 1968). These wide differences in the heats of combustion suggest that both the pulp yield and the amount and composition of wood organics recovered as by-products will influence the heat of combustion of the black liquor and the thermal efficiency of the recovery boiler.

Reduced sulfur compounds in black liquor also contribute to its heat of combustion. The heats of combustion of Na_2S and $Na_2S_2O_3$ are 12,900 and 5790 kJ/kg, respectively. Their contribution to the heat of combustion (obtained from bomb calorimetry) of black liquor is included only if the reduced sulfur compounds were not oxidized during drying of the liquor solids. Since sulfur compounds leave the recovery boiler in a reduced state, the energy required to reduce them must be taken into account. This correction is the heat of reduction of Na_2SO_4 (a bomb calorimeter product) to Na_2S (a recovery boiler product), adjusted by the degree of conversion of the reduction reaction.

The heating value of black liquor solids can be estimated from heats of combustion and mass balances around the pulp digester. The mass of organics in black liquor (M_{org}) per ton of dry pulp, and their average heat of combustion (HHV_{org}), are given by Equations 1 and 2, respectively:

$$M_{org} = \frac{1}{Y_{tot}} \sum_i W_j(1-Y_j) = \frac{1-Y_{tot}}{Y_{tot}} \quad (1)$$

Where

Y = mass fraction of wood components retained as pulp, dry wood basis,
W = mass fraction of wood components in dry wood,

tot = mass average of all components, and
j = component of wood

$$HHV_{org} = \frac{\sum_j W_j(1-Y_j)HHV_j}{(1-Y)_{tot}} \quad (2)$$

The mass of inorganic pulping chemicals per ton of pulp (M_i) and the heat of combustion of the reduced sulfur compounds

(HHV$_s$) per unit of Na$_2$S originally present in the pulping chemicals are given by Equations 3 and 4, respectively:

$$M_i = \frac{C_1 \times C_2}{Y_{tot}} \tag{3}$$

Where

C_1 = Chemical/wood ratio as active alkali, and

C_2 = total inorganics/active alkali ratio in pulping liquor

Active alkali is the sum of Na$_2$S and NaOH delivered to the digester as equivalent mass of sodium oxide (Na$_2$O).

$$HHV_s = .70 \; HHV_{Na_2S} + .296 \; HHV_{NA_2S_2O_3} \tag{4}$$

The average heating value for the black liquor solids (HHV$_{bls}$) is then calculated using Equation 5:

$$HHV_{bls} = \frac{M_{org}HHV_{org} + M_i X_{Na_2S} \; HHV_s}{M_{org} + M_i} \tag{5}$$

Where X = mass ratio of reduced sulfur compounds (as Na$_2$S equivalent) to inorganics in the liquor. Equation 5 provides a convenient method to estimate the heating value of black liquor. It is based on three assumptions: (1) there is no change in the heat of combustion of the wood organics during pulping; (2) all wood organics not recovered as pulp are recovered as black liquor solids; and (3) 30% of the Na$_2$S originally present in the pulping chemicals is oxidized to Na$_2$S$_2$O$_3$.

III. COGENERATION

Cogeneration may be accomplished by the generation of non-condensing power, expanding steam through a turbine before it is used for process heating. Figure 2 shows a simplified flow diagram for a cogeneration system. High pressure steam that is generated in the recovery and auxiliary boilers flows through a power turbine to generate electricity. The low pressure turbine exhaust is used for process heating in the pulp and paper mill. The ratio of electrical energy generated to low pressure steam flow depends on the turbine efficiency and the turbine inlet-outlet conditions. The boiler fuel

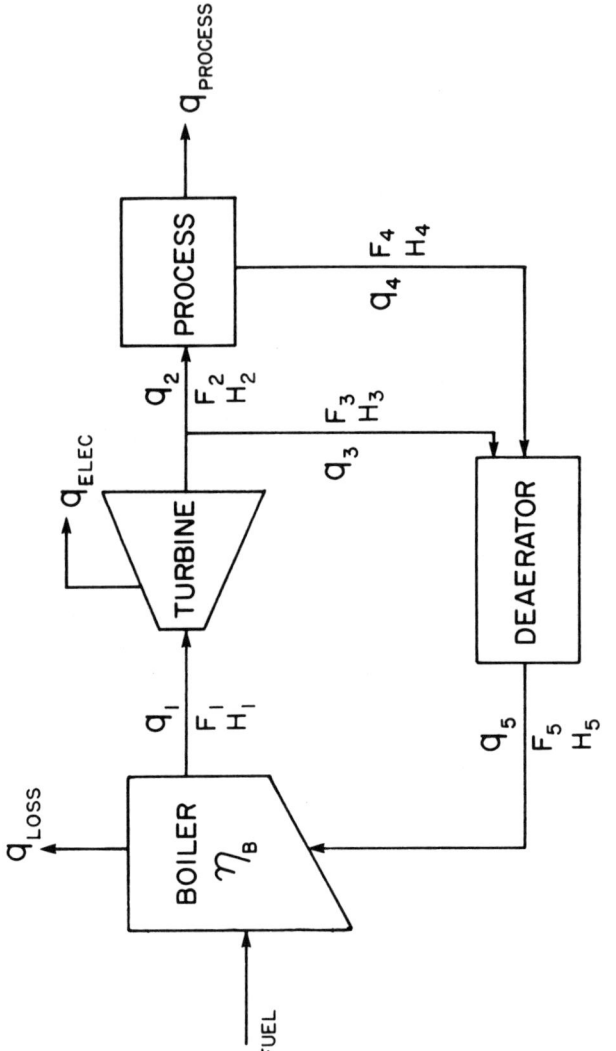

FIGURE 2. Cogeneration system of a pulp mill (q = energy flow rate, F = mass flow rate, H = specific enthalpy).

Energy Costs of Increased Organics Recovery

requirements and the rate of electrical energy generation can be calculated from the process heating requirements if the steam conditions around the cogeneration system and the boiler thermal efficiency are known. The calculation requires the simultaneous solution of mass and energy balances around the deaerator, boiler, and turbine. Referring to Figure 2, the mass flow of steam (F_2) or condensate (F_4) is calculated from the known rate of energy flow required for process heating ($q_{process}$), and the specific enthalpy of process steam (H_4):

$$F_2 = F_4 = q_{process}/(H_2-H_4) \tag{6}$$

The deaerator mass and energy balances must be solved simultaneously before the turbine exhaust flow is calculated:

$$F_5 = F_3 + F_4 \tag{7}$$

$$F_5 \text{x} H_5 = F_3 \text{x} H_3 + F_4 \text{x} H_4 \tag{8}$$

$$F_5 = F_1 = F_2 + F_3 \tag{9}$$

The rate of electrical power generation (q_{elec}) is calculated from the steam flow to the turbine (F_1) and the enthalpy change across it (H_1-H_2):

$$q_{elec} = F_1(H_1-H_2) \tag{10}$$

The fuel required (q_{fuel}) to meet the requirements for process heating of the mill are calculated by an overall energy balance:

$$q_{fuel} = (q_{elec} + q_{process})/E_b \tag{11}$$

An efficiency for conversion of the heat not used in process can be defined as (Kehlhofer, 1981):

$$E_{elec} = q_{elec}/(q_{fuel} - q_{process}/E_b) \tag{12}$$

There are two different procedures for attaching a value to electrical energy produced by a cogeneration system. In the first, the total cost of the fuel used to meet the requirements for process heating both with and without cogeneration is determined. The cost of generating electrical energy is the difference between the two. This method is most useful when designing a new mill or when making major changes in the cogeneration system. In the second method, the value of the cogenerated electricity is fixed by the replacement cost if that energy were purchased from a source outside the mill.

The value of the low pressure steam becomes the difference between the cost of producing high pressure steam and the cost of the electrical energy if it were purchased offsite. This method should be applied when minor changes in process operation or configuration are made that change the net flow of steam through the turbine.

IV. PULP MILL OPERATING PRACTICES AND ENERGY CONVERSION

The way in which a chemical pulp mill is operated is another important factor in determining the net energy recovered from kraft black liquor and the purchased energy requirements of a pulp and paper mill (Frederick, 1981). Pulp mill operating practices impact the energy recovered from combustion of black liquor in two ways. The first is related to black liquor composition and its relationship to heat of combustion. Any changes in process operation that increase the ratio of either inorganics or water to dissolved wood organics will decrease the thermal efficiency of the recovery boiler. It should be noted that any process changes that increase the inorganic/organic ratio in black liquor will also increase the water/organic ratio if the solids concentration of the liquor going to the boiler remains constant. For example, consider a pulp mill operating at the conditions shown in Table II. If the ratio of pulping chemicals/wood is increased by 10% and the rate of flow of organics to the boiler is constant, then the heat of combustion of the black liquor solids will be reduced by 3.5%. The amount of water input to the boiler with the black liquor will increase by 4.1% and the thermal efficiency of the recovery boiler will be reduced by 2.3%. Similar effects are seen when the ratio of inactive to active chemicals in the pulping liquor increases. These results correspond to a pulp mill operating as shown in Table II and a recovery boiler operating as shown in Table I. The important point here is that the energy flows in the cogeneration system can be altered significantly by small changes in process operation.

There is not a straightforward relationship between pulp yield, heat of combustion of black liquor solids, and thermal efficiency of the recovery boiler. An increase in pulp yield will increase heat of combustion and thermal efficiency if the yield increase is obtained at a constant pulping chemical/wood ratio. If the chemical/wood ratio is changed, then the heat of combustion can change in either direction, depending upon the resulting inorganic/organic ratio and the composition of the dissolved organics. The impact on energy recovery of pulp

TABLE II. Operating Conditions for a Typical Kraft Pulp Mill (Basis is 1 metric ton dry pulp)

Wood species pulped	softwoods
Chemical/dry wood mass ratio[a]	.18
Total inorganics/active alkali	1.67
Pulp yield (dry wood basis)	45%
Black liquor solids flow rate	1890 kg
Black liquor solids concentration to the recovery boiler	65%
Recovery boiler thermal efficiency	66.5%

[a] Chemical/wood ratio as active alkali is defined as the total mass of NaOH and Na_2S (or NaOH and NaHS) delivered to the digester, expressed as sodium oxide (Na_2O) equivalent, per unit of wood mass.

yield and chemical/wood ratio can be calculated using Equations 1-5.

The second factor influencing energy recovery is that changes in process configuration or mill operating practices usually change the energy required to operate the pulp mill. This does not directly impact energy balances around the recovery boiler, but does change the purchased energy requirements for the mill. For example, more steam is required to concentrate the spent pulping liquor when the pulping chemical/wood ratio is increased. This means that less net steam is available for other uses. Fuel requirements for calcining also increase when the chemical/wood ratio increases. The overall result is an increase in purchased fuel requirements for the mill. The magnitude of the changes in energy flows around the pulp mill is not easy to estimate by back-of-envelope calculations. It depends upon many pulp mill parameters and is best calculated by steady-state process simulation (Baldus and Edwards, 1979).

V. COST ANALYSIS OF ORGANICS RECOVERY

The energy costs of interest in operating a pulp and paper mill are for purchased fuel and electricity. The purchased energy requirements depend strongly on how the mill is configured and operated, and are sensitive to process modifications or changes in process operating conditions. For the recovery boiler, changes in process configuration or operating practices elsewhere in the pulp mill usually result in changes in the fuel input rate and thermal efficiency. Since these both impact the rate of energy recovered as steam, they also impact purchased energy requirements. The rate of fuel input to the recovery boiler can be determined from mass balances around the mill. Thermal efficiency depends on the composition of the fuel and how it changes. The incremental thermal efficiencies for the components of black liquor differ depending on their elemental composition and heat of combustion. For example, black liquor containing 65% solids may be burned to generate steam with a thermal efficiency of 66.5% (Table I). The dissolved lignin and carbohydrates from the liquor burn separately with thermal efficiencies of 82.5% and 77.4%, respectively, when burned at the same solids concentration and boiler operating conditions.

When organics are removed as by-products from black liquor, fuel must be purchased to generate the same amount of steam. We can define the minimum value of the organics recovered as the cost of their recovery plus the cost of replacing them with purchased fuel equivalent to their steam generating capacity. The cost of recovery depends strongly on the separation technology employed. Only the cost of fuel replacement will be considered here. Since the fuel costs depend on the operating parameters of the pulp mill, the calculations made in the rest of this section are based on the conditions given in Table II.

There are two principal constraints on replacing black liquor with another fuel: (1) the mill demand for steam must be met; and (2) the replacement fuel must be compatible with the available combustion facilities. The choices of replacement fuels for most pulp mills are hog fuel, coal, oil, or natural gas. If there is excess capacity available for burning hog fuel or coal in an auxiliary boiler, then the steam generation rate of the recovery boiler can be reduced as wood organics are removed from the black liquor. However, if the recovery boiler must be used fully to meet the steam requirements of the mill, then oil or natural gas must be burned in it to replace the wood organics.

The actual cost of replacement fuels depends on the combustion properties of the fuels as well as their unit costs. We can choose oil and hog fuel as the fuels which bound the range of costs of replacement fuels; coal and natural gas fall between these two. In the following calculations we assume the following combustion properties for these fuels: hog fuel has a heat of combustion of 20,000 kJ/kg (dry basis) and burns with a thermal efficiency of 68% when it contains 50%

Energy Costs of Increased Organics Recovery

moisture; oil has a heat of combustion of 42,000 kJ/kg and burns with a thermal efficiency of 87%.

The fraction of the dissolved organics that can be recovered from black liquor is determined by the liquor's ability to sustain stable combustion, which in turn depends on its heat of combustion. Because the composition of the organics determines the heat of combustion of black liquor, as they are removed the heat of combustion decreases. The heat of combustion of black liquor solids (HHV_r) can be calculated as a function of the quantity of organics removed by applying mass and energy balances around the point at which the organics are removed. Equations 13-15 give the results:

$$M_o = 1000 \,(((1/Y_{tot}) - 1) + C1 \mathrm{x} C2/Y_{tot}) \tag{13}$$

$$M_i = 1000 \,(Z_i/Y_{tot}) \tag{14}$$

$$HHV_r = \frac{M_o \mathrm{x} HHV_o - \Sigma M_i \mathrm{x} HHV_i}{M_o - M_i} \tag{15}$$

Where

HHV_o = heating value of black liquor solids before organics are removed,

M_o = mass of black liquor solids per metric ton of dry pulp,

M_i = mass of component i of organics removed per metric ton of dry pulp, and

Z_i = mass of organics removed from black liquor as a fraction of dry wood mass before pulping.

Using the values of HHV_o from Table I and those for Y_{tot} and HHV_i specified earlier in this chapter, Equation 15 reduces to Equation 16 for lignin and carbohydrate removal:

$$HHV_r = \frac{15300 - (31000 \mathrm{x} Z_L) - (20200 \mathrm{x} Z_C)}{1 - (1.18 \mathrm{x} Z_L)} \tag{16}$$

Where

Z_L = lignin removed, mass fraction of dry wood weight before pulping, and

Z_C = carbohydrate removed, mass fraction of dry wood weight before pulping.

There are two bounds to the impact of recovering the major organic constituents from black liquor. One is removal of lignin only, which decreases the heat of combustion of black liquor solids most rapidly for a given quantity of wood organics removed. The other is removal of carbohydrates only, which decreases the heat of combustion most slowly per unit of wood organics removed. The effect of removing a lignin-carbohydrate mixture falls between the two.

Figure 3 shows the relationship between the heat of combustion of the black liquor and the fraction of recovered wood organics. It is based on Equation 16. Using 12,600 kJ/kg as the combustion stability limit, then 16.6% of the wood (based on input to the digester) can be removed if only lignin is recovered. If carbohydrate is recovered, the energy balance analysis indicates that the limit is 50%, which exceeds the quantity of carbohydrates in black liquor. The actual limit is about 27%, corresponding to all of the carbohydrate in the dissolved wood organics. These limits are quite sensitive to the ratio of organics/inorganics in black liquor. Also, they apply only when no additional fuel is burned in the recovery boiler. However, if additional fuel is burned there, the average heat of combustion of all streams entering the recovery boiler must still equal or exceed 12,600 kJ/kg.

The rate of steam generation in the recovery boiler when wood organics are removed from black liquor is the difference between: (1) the steam generating capacity of the boiler when fired with black liquor before organics are removed, and (2) its capacity if it were fired with the recovered organics at the same solids content:

$$Q_{stm} = M_o \times HHV_o \times E_b - M_i \times HHV_i \times E_i \tag{17}$$

Where E_i equals the incremental thermal efficiency for combustion of the organics removed from black liquor.

Figure 4 shows how the steam generating capacity of the recovery boiler decreases as dissolved wood organics are removed from the black liquor. At the removal limits for lignin or carbohydrates, the steam generating capacity of the recovery boiler is reduced from 18.9 million kJ/metric ton of pulp to 11.7 million kJ/metric ton. This corresponds to a reduction of about 38% and is equivalent to roughly 440,000 barrels of oil for 1000 metric ton/day pulp mill.

The cost of the replacement fuel ($F) is calculated from the change in energy recovered as steam when organics are recovered (ΔQ_{stm}) (Figure 4), the unit cost of replacement fuel (C_{fuel}), and the thermal efficiency of the boiler (E_b) in which the replacement fuel is burned:

$$\$F = (Q_{stm} \times C_{fuel})/E_b \tag{18}$$

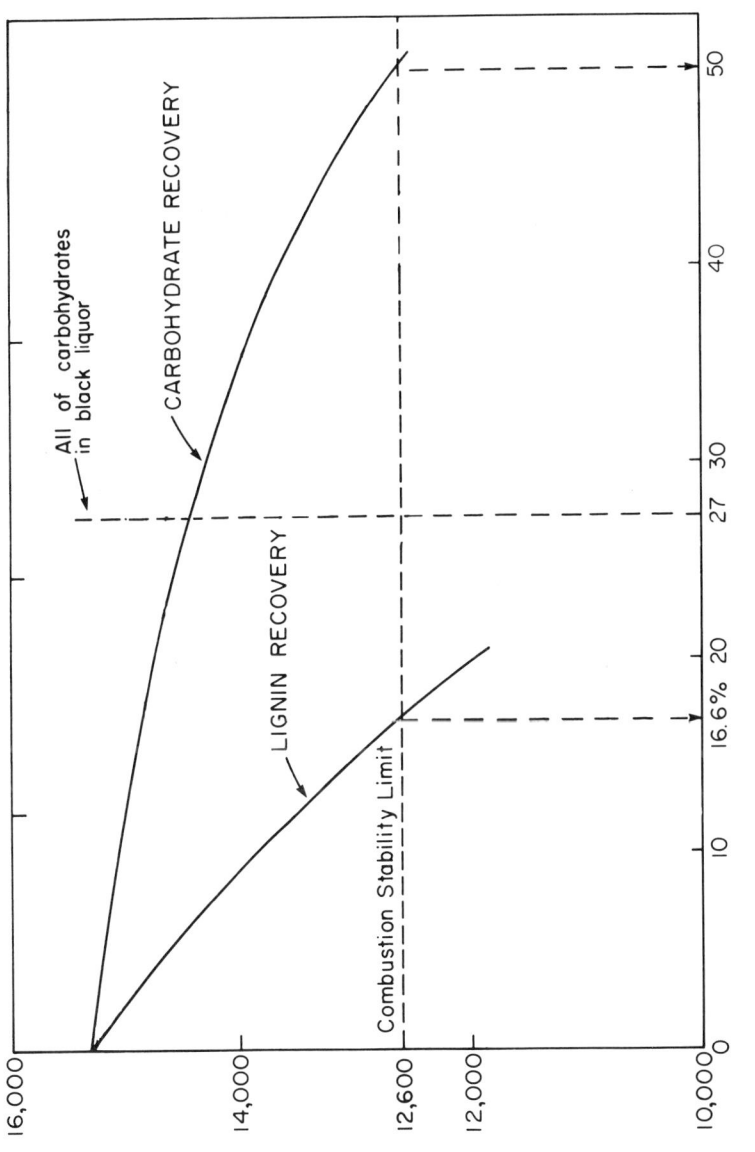

FIGURE 3. Relationship between the heating value of black liquor solids and the amount of organics recovered as by-product chemicals.

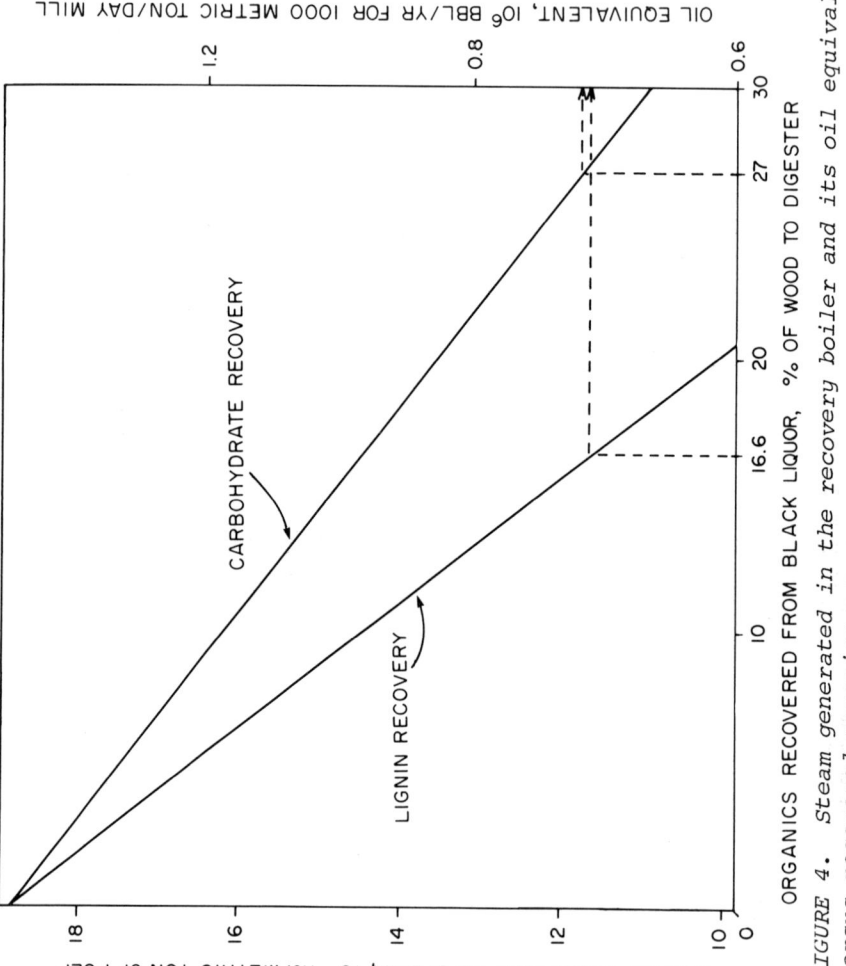

FIGURE 4. Steam generated in the recovery boiler and its oil equivalent versus recovered organics.

Figure 5 shows how the annual replacement cost for fuel varies with the type of replacement fuel used, its unit cost, and the chemical composition of the wood organics recovered. At the same fuel unit cost, the annual cost of replacement with hog fuel is higher because of its lower thermal efficiency. However, hog fuel generally costs less than oil, differing by as much as a factor of 5 (Jamison, 1979). At the removal limit established by combustion stability and mass balance constraints, there is no effect on annual replacement fuel cost from the composition of the organics removed.

The cost increment for the fuel replacing the recovered wood organics is shown in Figure 6. It is calculated by dividing the cost of the replacement fuel by the quantity of wood organics removed from black liquor. The fuel replacement cost is about $.08/kg carbohydrate or $.13/kg lignin if oil at $5/million kJ is used as replacement fuel. Using hog fuel at $2/million kJ, the costs are $.035/kg for carbohydrates and $.06/kg for lignin.

VI. SUMMARY AND CONCLUSIONS

It has been shown that the relationship between pulping chemistry, pulp mill operating parameters, and a mill's purchased energy requirements are complex. The heat of combustion of black liquor solids, thermal efficiency of the recovery boiler, and a mill's purchased energy requirements can all be affected by changes in the pulping and recovery processes.

When lignin or carbohydrates are recovered as by-products from black liquor, their removal is limited by its impact on combustion stability in the recovery boiler. The cost of replacing the by-product wood organics with another fuel depends on the cost of the fuel and the thermal efficiencies associated with burning both the replacement fuel and the organic component recovered from black liquor. At conventional fuel prices, the range of replacement fuel costs is $.035 - .13/kg organics recovered.

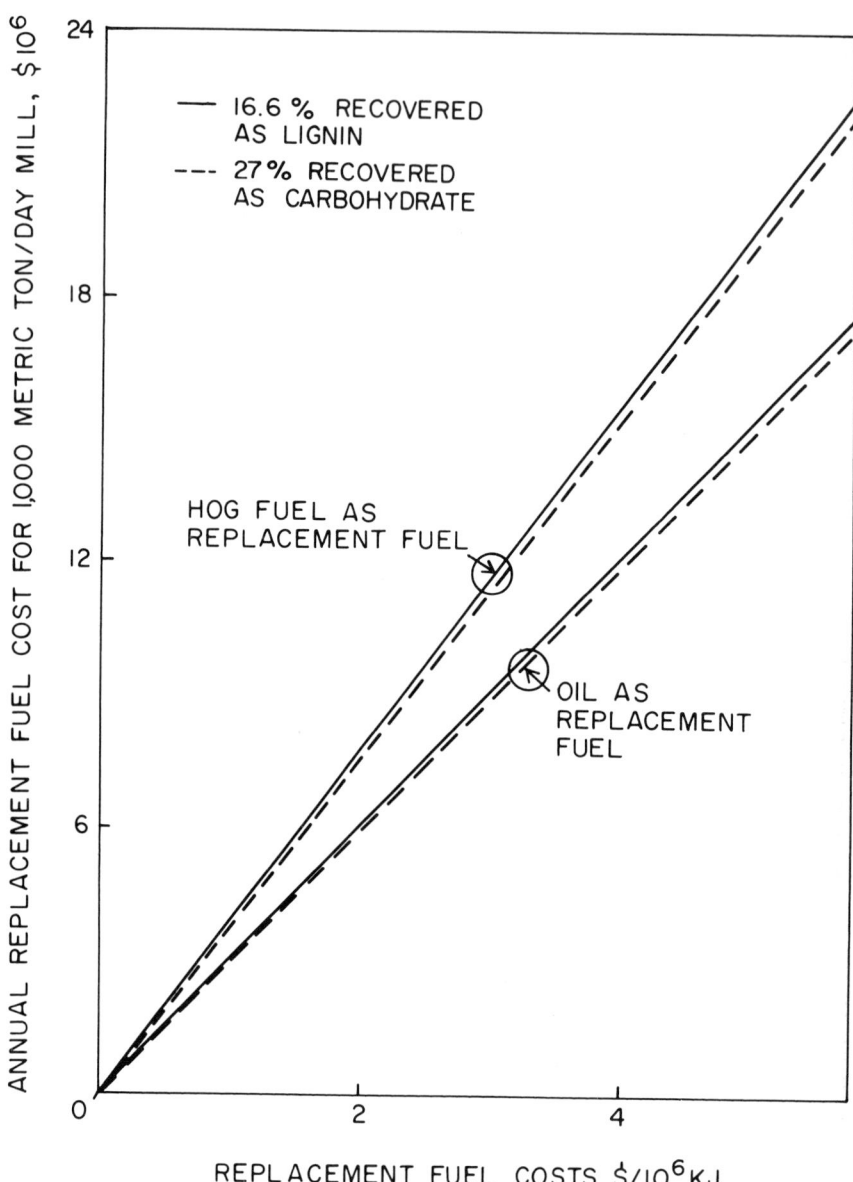

FIGURE 5. Annual fuel replacement costs as a function of unit fuel cost, fuel type, and fraction of organics recovered.

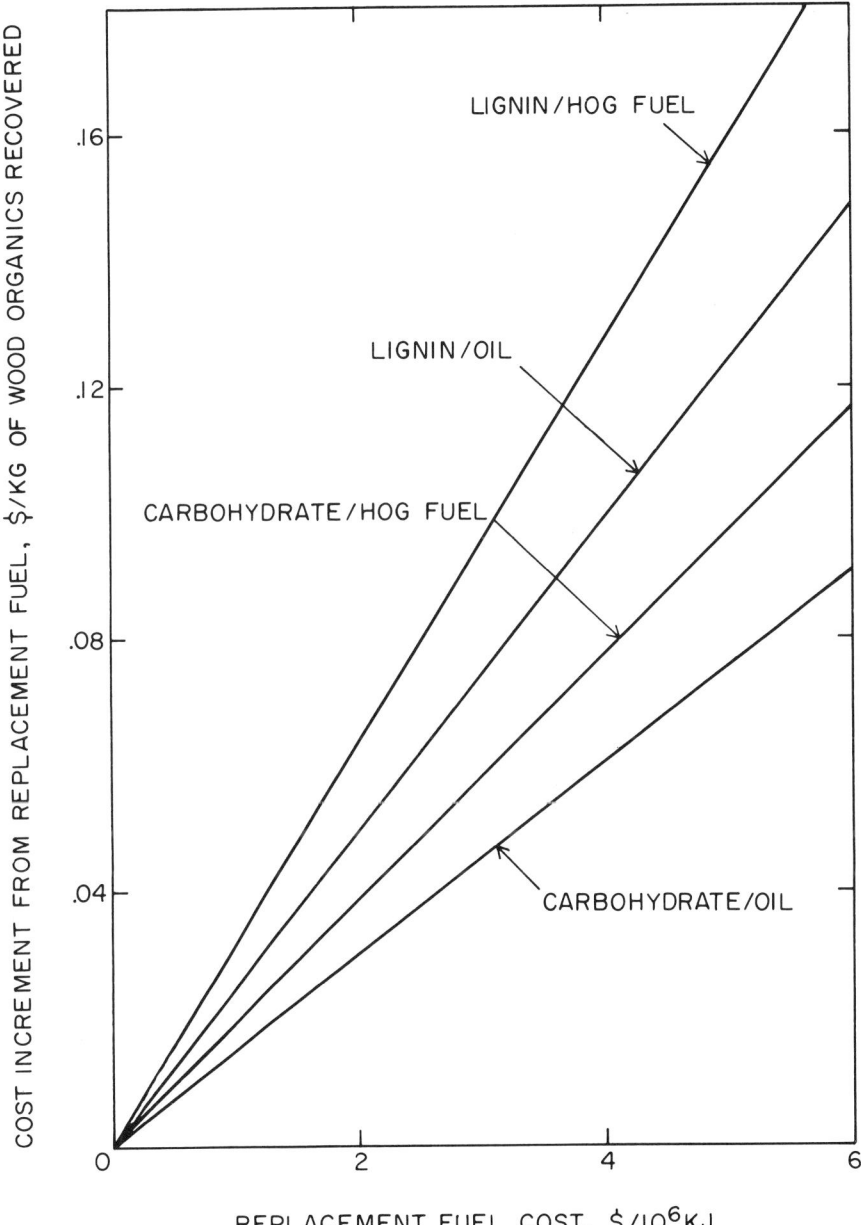

FIGURE 6. Incremental cost of recovered organics versus unit cost of replacement fuel.

REFERENCES

Annergren, G. E., Haglund, A., and Rydholm, S. A. 1968. *Svensk Paperstidning* 71(15):497-504.

Baldus, R. F., and Edwards, L. L. 1979. "Mass and Energy Balances for Complete Kraft Mills." AIChE Symp. Ser. 184, Vol. 75.

Casey, J. P. 1980. "Pulp and Paper Chemistry and Chemical Technology," Third Edition, Vol. 1. Wiley & Sons, New York.

Frederick, W. J. 1981. Weyerhaeuser Approach to Energy Audits for the Kraft Pulping and Recovery System. "Proceedings of the International Conference on Recovery of Pulping Chemicals," Vancouver, B.C.

Grant, T. J., and Slinn, R. J. 1983. "Patterns of Fuel and Energy Consumption on the U.S. Pulp and Paper Industry." The American Paper Institute, New York.

Jamison, R. L. 1979. Wood Fuel Use in the Forest Products Industry. *In* "Progress in Biomass Conversion" (K. V. Sarkanen and D. A. Tillman, eds.). Academic Press, New York.

Kehlhofer, R. 1981. *Combustion*. March, p. 22.

Reeve, D. W., Pryke, D. C., Lukes, J. A., Donovan, D. A., Valiquette, G., and Yemchuk, E. M. 1983. *Pulp and Paper Canada* 84(1):T25-29.

Rydholm, S. 1965. "Pulping Processes." Interscience Publishers, New York.

TAPPI. 1979. "Recovery Boiler Test Procedure Short Form," TIS 407-22, Technical Association of the Pulp and Paper Industry, Atlanta, GA.

Tran, H. N., Reeve, D. W., and Barham, D. 1983. *Pulp and Paper Canada* 84(1):T7-12.

MICROECONOMIC APPROACHES
TO BIOMASS FUEL PRICING

E. C. Lesnick, Jr.

EBASCO Business Consulting Company
A Division of EBASCO Services, Inc.
Mountain View, California

I. INTRODUCTION 164

II. NATURE OF BIOMASS FUELS 164

III. MICROECONOMIC THEORY AND BIOMASS FUEL PRICING . . . 165

    A. Basic Considerations 167

    B. Supply and Demand for Biomass Fuels 167

    C. Transfer Pricing 179

IV. APPLICATION OF MICROECONOMIC CONCEPTS 185

    A. Supply and Demand Analysis for Wood Fuel . . . 185

    B. An Alternative Cost Approach to Value
       Wood Fuel 189

    C. Transfer Pricing: A Linear Programming
       Approach 198

V. CONCLUSIONS 201

I. INTRODUCTION

The economic framework for price determination is the marketplace and the supply and demand relationships which exist in competitive and imperfect resource markets. If no market exists, there is no market price which can be used to value the biomass resource. In this situation, the resource value may still be determined by imputation or market simulation methods. These fundamental economic considerations are elaborated upon further in the following sections.

Biomass fuels encompass several distinct types of resources each of which should be considered in light of market and nonmarket factors which affect the pricing decision. Biomass fuels include wood waste products, municipal refuse, walnut shells, and a number of other related products which are sold or transferred in various types of market situations. Factors such as market structure, spatial configuration, producer and consumer behavioral demand, and the role of government are important because they are instrumental in affecting the production and consumption of biomass resources and consequently their associated prices.

II. NATURE OF BIOMASS FUELS

By their nature, biomass resources are not homogeneous fuel resources like electricity, water, or even the various fossil fuels. Instead, biomass resources are heterogeneous in nature consisting of a wide range of renewable energy products (see chapter by A. J. Rossi). Biomass resources are not determined solely by "mother nature" but by man and society at large through their economic and industrial activities. Also, biomass fuels tend to be widely dispersed over most land areas and not as geographically concentrated as coal, oil, and natural gas. Further, some biomass resources (e.g., wood) can be stored for future use similar to other fuels but unlike electricity.

The multiple uses of biomass resources must be recognized in economic analyses. The supply and utilization of biomass fuels are affected by the suitability and cost of biomass resources to satisfy other nonfuel competing household or industrial needs. Also, biomass fuel prices should reflect the value that such resources command in alternative sources of employment; that is, their opportunity costs.

A summary of the various classes or types of biomass resources from which biomass fuels can be derived is presented below based on a recent study (Benemann, 1980).

There are four basic types of biomass resources:
(1) waste products, (2) residues, (3) energy crops, and
(4) integrated biomass systems.

Wastes encompass organic materials that are generated as a negative byproduct of consumptive or productive economic activity. In general, wastes accumulate at the site of such activities and must be disposed of so as not to impede further activities. These economic activities might produce municipal solid wastes, some forest wastes, and animal manures. There are collection, storage, transportation and other types of costs associated with the handling and processing of waste material.

Residues represent plant materials left behind after productive activity such as harvesting crops or sawing timber. These products have no production costs but may have collection storage and disposal costs associated with them. Further, a purchase cost must be paid to the resource owner because these residues usually have some positive commodity value or opportunity cost.

Energy crops are grown specifically for their calorific value or fuel content. These crops include short rotation tree farms, ocean kelp farms, land based aquatic plant systems, or sweet sorghum. The value of any fuels produced would need to exceed the costs of all factors used in production, including land and related harvesting/collection costs to be economic.

Integrated biomass systems involve multiple products with fuel as one of the products. For example, sugar products like molasses or corn substrates could be used as fermentation feedstocks to produce alcohol.

Table I contains a summary of biomass resources in the United States based on data sources published in the years 1977-1979. For each type of resource, the total quantity of the biomass resource is shown as well as the estimated amount that could be recovered for fuel and the fuel equivalency measure in quadrillion Btu (quads). It appears from the data in Table I that residue wastes from cultivated crops and commercial forests, and municipal solid wastes are the most promising biomass fuel sources based on present technology and economic conditions. However, in future years, noncommercial forestry and short rotation tree farms may also provide significant fuel sources.

III. MICROECONOMIC THEORY AND BIOMASS FUEL PRICING

This section presents the fundamental economic theory underlying the pricing of biomass fuels. The theoretical

TABLE I. Biomass Resources in the United States[a]

Type of resource	Total biomass resource (10^6 tons/yr)	Estimated recoverable for fuel (10^6 tons/yr)	Fuel equivalency[b] (Quads)
Cultivated crops (residue wastes)	300	100	1.0
Commercial forests (residue wastes)	200	100	1.0
Animal husbandry (manures)	220	26	0.2
Municipal solid wastes	150	100	1.0
Municipal liquid wastes	13	13 (yr. 2020)	0.1
Noncommercial forestry	large	400 (yr. 2020)	4.0
Sugar crops for gasohol	medium	200 (yr. 2000)	0.5
Short rotation tree farms	large	500 (yr. 2020)	5.0
Ocean farming	large	Uncertain, probably very low	-0-

[a] Source: Benemann, 1980.
[b] Fuel equivalency assumes that 10×10^6 Btu/ton of dry biomass and 6×10^6 Btu/ton of wet biomass are recoverable as fuels.

principles are developed gradually, first distinguishing between consumer and producer behavior and then examining the determination of fuel prices in various markets by the forces of supply and demand. Also, the notion of transfer pricing is analyzed. These theoretical principles should be well known by resource economists and economists in general and consequently this section may offer only a review of basics. However, for many engineers, resource planners, or other professional persons the theoretical concepts presented may be less well known and understood. This section is intended for the noneconomist to enhance their understanding of biomass resource pricing decisions.

Microeconomic Approaches to Biomass Fuel Pricing

A. Basic Considerations

Biomass fuel prices are determined by the forces of supply and demand in highly competitive or imperfectly competitive biomass resource markets. These markets can be chiefly characterized by the type of biomass fuel, number and type of buyers and sellers, government rules and regulations, and physical location. Purely competitive resource markets would imply that neither individual resource suppliers nor demanders could influence price by their actions in the marketplace. Also, variable biomass resources are free or mobile between alternative employment opportunities and market prices are flexible. In imperfectly competitive markets, one or more of these conditions do not hold.

Biomass fuels are considered to be intermediate type commodities or inputs into the production of some other good which is ultimately demanded. As such, the demand for biomass fuels is considered to be a derived demand. For example, hogged bark from a sawmill does not normally have a market value as a final good but it may be in demand for the production of steam and/or electricity if economic conditions are favorable.

Further, biomass fuels are demanded not only by producers in the business sector, but consumers as well in the industrial and household sectors. For example, wood wastes can be used by households for heating or cooking purposes or by manufacturers to produce steam. The implication of this is that the consumption behavior of both producers and consumers with respect to the specific biomass fuel should be taken into consideration in performing any economic analysis.

B. Supply and Demand for Biomass Fuels

1. Household Demand. According to standard neoclassical economic analysis, household consumers want to maximize utility subject to a budget constraint (Ferguson, 1972). Let us assume for simplicity, but without loss of generality, three goods:

Q_1 - fuel type 1; Q_2 - fuel type 2; Q_3 - nonenergy good with corresponding prices P_1, P_2 and P_3.

The consumer has a given money income (Y) and his utility function is represented by $U = U(Q_1, Q_1, Q_3)$; the budget constraint is:

$$Y = P_1Q_1 + P_2Q_2 + P_3Q_3 = \sum_{i=1}^{3} P_iQ_i \qquad (1)$$

To maximize (U) subject to the given budget constraints, form the lagrangean (L):

$$L = U(X_1, X_2, X_3) - \lambda \left(\sum_{i=1}^{3} Q_1 P_1 - Y \right) \qquad (2)$$

where lambda (λ) is the lagrangean multiplier.

The first order conditions for optimization are:

$$\frac{\delta L}{\delta Q_1} = \frac{\delta U}{\delta Q_1} - \lambda P_1 = 0 \qquad (3)$$

$$\frac{\delta L}{\delta Q_2} = \frac{\delta U}{\delta Q_2} - \lambda P_2 = 0 \qquad (4)$$

$$\frac{\delta L}{\delta Q_3} = \frac{\delta U}{\delta Q_3} - \lambda P_3 = 0 \qquad (5)$$

$$\frac{\delta L}{\delta \lambda} = \Sigma_i Q_i P_i - Y = 0 \qquad (6)$$

The solution of the four equations with four unknown Q and λ, if it exists, is the system of demand functions:

$$Q_i = F_i(Y, P) \qquad (7)$$

Demand functions of the general form shown by equation (7) are thus assumed to exist. These relationships show quantities demanded of good i as a function of prices and total money income. For our analyses, it is assumed that total expenditures on all goods i exhaust money income and thus the terms are used interchangeably.

The implications of a linear budget constraint for consumer demand have been discussed in the literature (Deaton and Muellbauer, 1980) and are summarized in the following discussion. Since the demand functions satisfy the budget constraint, a constraint on the function F_i is implied as follows:

$$\sum_i P_i F_i(P, Y) = Y \qquad (8)$$

This constraint is termed the adding-up restriction. Another implication of the budget constraint is the homogeneity restriction. This restriction suggests that the demand functions are homogeneous of degree zero:

$$F_i(\lambda Y, \lambda P) = \lambda° F_i(P,Y) = F_i(P,Y) \tag{9}$$

The adding-up restriction implies that for all i, i=1,....n

$$\sum_k P_k \frac{\delta F_k}{\delta Y} = 1 \tag{10}$$

$$\text{and} \sum_k P_k \frac{\delta F_k}{\delta P_i} + Q_i = 0 \tag{11}$$

This allows for changes in Y and P resulting in rearrangements in purchases which still do not violate the budget constraint. The first part (10) above is called Engel aggregation and the latter (11), Cournot aggregation.

The homogeneity restriction implies that for all i=1,....n

$$\sum_k P_k \frac{\delta F_i}{\delta P_k} + Y \frac{\delta F_i}{\delta Y} = 0 \tag{12}$$

This means simply that a proportionate change in P and Y will leave purchases of good i unchanged.

It is useful to introduce the term budget share at this point. The budget shares (w_i), defined by $w_i = P_i Q_i/Y$, represent the fraction of total expenditures allocated to each good. The logarithmic derivatives of the demand functions determine the total expenditure elasticities and price elasticities:

Total Expenditure Elasticity:

$$\eta_i = \delta\log F_i(Y,P)/\delta\log Y \text{ where } i=1,n \tag{13}$$

Price Elasticity:

$$\eta_{ij} = \delta\log F_i(Y,P)/\delta\log P_j \text{ where } i,j=1,n \tag{14}$$

The diagonal elements (η_{ii}) are the own price elasticities and the off diagonal elements (η_{ij}) are cross price elasticities.

Price and income elasticities of demand for fuel products are of considerable practical importance in exploring changes in the consumption of fuel due to changes in the prices of substitute and complementary products, and household income.

Price elasticity of demand measures the relative responsiveness of quantity demanded to changes in price. As indicated above, there are two types of price elasticities: (1) "own" price eleasticity and (2) cross price elasticity. Own price elasticity is determined by the percentage change in the quantity demanded of a good relative to a one percent change in the price of that same good. Cross price elasticities of demand measure the effect of a change in the price of one good on the consumption of another good. Income elasticity of demand measures the percentage change in quantity demanded relative to the percentage change in income. The formulae for income elasticity and price elasticity are shown in equations (13) and (14).

The rising cost of electricity in the United States has influenced many households to switch to alternative fuels such as natural gas and wood fuel for space heating and cooking purposes. Improvements in wood burning technologies have reduced the costs of supplying thermal energy in homes. The apparent upward trend in the purchase of wood burning stoves as well as other heating/cooking equipment can be explained to a large extent by positive increments in money income or household budgets and fuel cost savings.

   2. Producer Demand. The business sector consisting of industrial and commercial suppliers of various products account for a greater demand for biomass fuels than the household sector. Biomass fuels are used to generate process steam or electricity or as in the case of cogeneration facilities both of these products. The biomass fuels could be hogged bark, sawdust, planer shavings, walnut shells, or a number of other residue types. Biomass fuels are an input to the productive process along with capital, labor, and other factors. The assumption is made that producers employ fuel and other factors of production in an efficient manner such that profits are maximized.

In the simple case of a single productive service, the individual demand curve for a productive resource is given by the value of the marginal product curve of the productive resource. The value of the marginal product is defined as the marginal product of the resource multiplied by the market price of the commodity or output in question. The use of a single variable resource in product is not very realistic. In general, there are a number of variable resources let alone alternative fuel types that may be used in the production of commodities. However, it is preferable to begin the theoretical discussion with the limited case of one variable input and then expand to the case of multiple variable resources.

The production function can be written as follows for one variable output:

$$Q = f(x) \tag{15}$$

Where Q = physical output X = quantity of variable input, and $f'(x)$ = marginal product of X.

If we assume that the producer is a perfect competitor in commodity and factor markets, the market prices of the commodity (P_Q) and the factor P_x are given. The profit function can be expressed as:

$$\pi = P_Q Q - P_x X - F = P_Q f(x) - P_x X - F \tag{16}$$

Where F is total fixed costs.

The producer adjusts input usage to maximize profits, i.e. in mathematical terms the first derivative is set equal to zero:

$$\frac{d\pi}{dx} = P_Q f'(x) - P_x = 0 \tag{17}$$

Rearranging,

$$P_Q f'(x) = P_x \tag{18}$$

The above expression denotes that the value of the marginal product (VMP) should equal the price of the variable input factor to attain a maximization of profit. This result is illustrated in Figure 1(a) where the horizontal line denotes that there is in effect an infinite supply of the factor available to the producer at the market determined price P. The demand curve for the variable input is given by the VMP_x curve. The optimal input usage would be X_1 where VMP_x intersects the supply curve (S_x).

In the more realistic case of several variable productive resource inputs, the demand curve of a productive resource is not given by the value of the marginal product. This is due to the interdependence of the inputs used in the production process. A change in the price of one input results in changes in the rates of utilization of the other inputs.

The situation is illustrated in Figure 1(b) for an outward shift of the value of marginal product curve. Assume initially that the price of the productive resource is Px_1, and the value of marginal product is represented by VMP_1, resulting in input usage X_1 at point A. If the price of the variable input Px decreases from Px_1 to Px_2, the quantity demanded of factor x would not be determined by Px_2 intersecting VMP_1 at point C. The reason is that the input ratios will be changed due to the change in the price of factor x. Since factor x is less expensive than before, more of it will be used in the production process, and it will be substituted

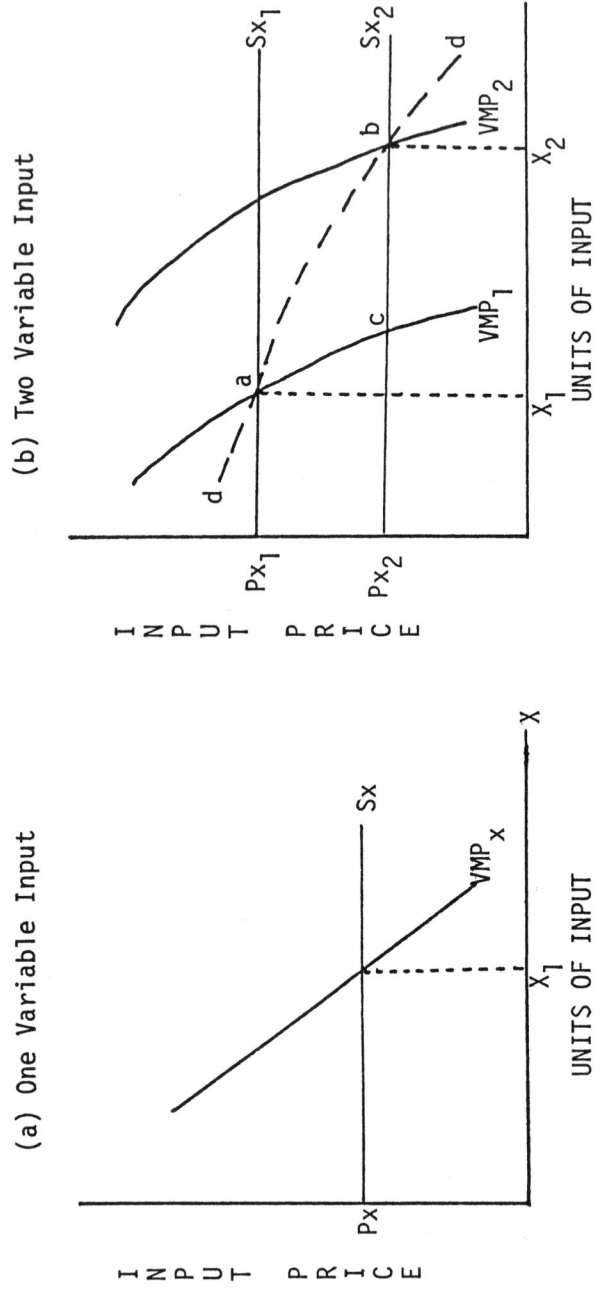

(c) Determination of Market Demand

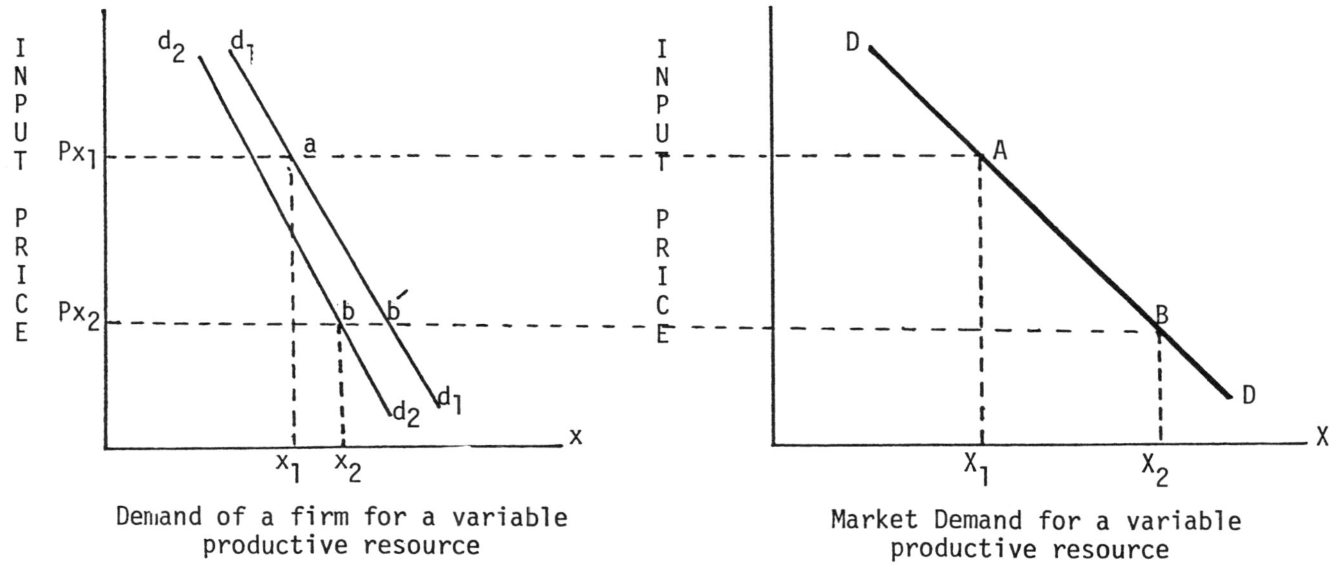

FIGURE 1. Demand for productive resource.

for the other variable inputs causing the marginal product curve of x to shift to the left. However as the quantity of x increases and output expands the usage of other inputs changes. This tends to shift the marginal product curve of x upward or to the right. In Figure 1(b), the VMP curve has shifted from VMP_1 to VMP_2 as a consequence of substitution, output and profit maximizing effects.

In summary, there are four main determinants of the demand for a productive resource. These determinants are as follows:

(1) The greater the quantity of other productive resources used in production, the greater the demand for a given quantity of the variable productive resource of interest.

(2) The demand price for a variable productive resource will be greater the higher the selling price of the commodity it is used to produce.

(3) The demand price for a variable productive service will be lower the greater the quantity of the resource in use.

(4) Changes in technology can make a variable input more productive and also makes the demand for any given quantity of it greater and vice versa.

Thus far, we have analyzed the demand for a productive resource in terms of an individual producer or household. In the next section, the aggregate or market demand for a variable productive resource is discussed and analyzed.

3. Market Demand. The market demand for a productive resource is the aggregate of all individual demand schedules. The process of aggregation is not simply the horizontal summation of input demands at each and every price of variable input. The process is illustrated in Figure 1(c) for a typical producer demand curve.

In the left panel of Figure 1(c), the typical producer demand curve is represented by $d_1 d_1$. At a price Px_1, the quantity of input X employed would be x_1 and market demand would be X_1 which would be determined by aggregating the quantities employed at price Px_1 over all producers. If the current price changes from Px_1 to Px_2, the quantity demanded would probably not be determined by moving down the curve $d_1 d_1$ to point b. When all firms increase their employment of factor X in production, total output of the commodity should expand leading to a decrease in the price of the commodity. The decrease in commodity price will result in a decline in value of marginal product. This is illustrated in the left graph in Figure 1(c) by the downward shift of variable input demand from $d_1 d_1$ to $d_2 d_2$. The quantity employed by the firm at a price Px_1 would be determined at point b on demand curve

d_2d_2; the amount is x_2. The market demand at input price P_x would be the aggregate of x_2 units of the productive resource over all resources employed. The point B represents the market demand at Px_2 price with X_2 units of the productive resource employed. If market demand curve is linear, the market demand curve could be represented by DD in the right graph shown in Figure 1(c).

4. *Supply of Biomass Fuels.* Biomass fuels can be treated in economic analysis as variable productive resources. These productive resources can be classified as natural resources or as intermediate goods. For example, residue materials and energy crops can be considered natural resources whereas residuals such as hog fuel are intermediate in nature. Irrespective of type, however, the supply curves for these resources are positively related to price; i.e., the supply curve is upward sloping. The basic economic principles behind the development of the supply relationship are elaborated upon in this section.

In general the production of variable productive resources can only be accomplished at increasing costs to the producer. This means that a producer would be willing to provide additional quantities of the resource if incremental costs could be recovered in the selling price or use value of the resource. This fundamental concept is developed below by analyzing the relationship between cost and resource output levels.

Short Run Cost Function: $C = F + V(Q)$

Where C = total costs
F = level of fixed costs
$V(Q)$ = variable costs

The average costs of production (AC) would be determined by (C/Q). Marginal Cost (MC) is the increment to total cost due to the addition of one unit of output. Marginal cost is equal to the derivative of the total cost function with respect to quantity of the variable productive resource.

In the short run, the cost function of producing or supplying biomass resources can be summarized by aggregating the fixed and variable cost components involved in the process. The fixed costs pertain to rental payments on land or leased plant and equipment, interest on borrowed capital, property taxes, and other expenses not related to changes in output. Variable costs account for the remaining short run costs. Variable costs include all costs that vary with the level of resource output. In general, the two main categories of variable costs for biomass fuel production would be: (1) collection, storage, and processing costs associated with the organic biomass resource, and (2) conversion of the raw biomass resource to other biofuels. The relevant variable costs would depend on the specific biomass resource or fuel under consideration.

Biomass resources similar to other energy resources such as coal and oil are mainly supplied by business firms or producers. The economic analysis behind the supply of these inputs given their demand is identically the same as that for the determination of the level of output in the theory of the firm. Therefore, it is assumed that producers or suppliers of biomass resources want to maximize profits. But the level of revenue to the firm facing a fixed price (P) for the biomass resource depends upon the level of biomass resource output.

The profit function can be written:

$$\pi = PQ - (F + V(Q)) \tag{19}$$

The first and second order conditions for profit maximization are as follows:

(a) First-order condition

$$P - \frac{d\,V(Q)}{dQ} = 0; \quad \text{or} \quad P = \frac{d\,VQ}{dQ} = V'(Q) \tag{20}$$

(b) Second-order condition

$$\frac{d^2\pi}{dQ^2} = \frac{-d^2C}{dQ^2} < 0; \quad \text{or} \quad \frac{d^2C}{dQ^2} > 0 \tag{21}$$

The first-order condition requires the producer to equate marginal cost with the fixed selling price of his biomass resource. Thus additional units of output would be supplied if incremental revenue exceeds incremental cost. The second-order condition calls for marginal cost to be increasing at the profit maximizing output level.

The market supply function for the biomass resource or input is the aggregate supply function of all firms which produce the biomass resource in question. In theory, the market supply function is positively related to price and therefore the supply curve slopes upward. The shape of the market supply curve for a biomass fuel is a subject for empirical estimation. For some resource markets, the supply curve might be relatively flat (highly price elastic) or upward sloping (price elastic) or vertical (totally price inelastic) at some level of output.

Resource owners or producers of biomass resources are influenced by the price of biomass resources. If the price or value in use of a given biomass resource rises due to an increase in demand relative to supply or a reduction in supply relative to demand, producers or biomass resource owners would be expected to increase their production or supply of the resource in the next supply period. The decision to increase supply would entail a flow of resources such as land, labor (person-days), capital (equipment and plant facilities), and ingredient type inputs such as fertilizer, water, etc. Some of these resources can be altered in the short run but other productive resources could only be changed in amount in the long run because they may require investments in time and money. Net investment in plant facilities and purchasing new acreage in land provide possible examples of resources that could be changed only in the long run.

a. *The effect of transport costs on supply.* Transportation costs associated with supplying biomass resources can have a significant impact on the supply of the resource. It is convenient to treat transportation costs for analysis purposes as proportional to distance eschewing the specific cost elements which must be evaluated in an actual situation.

The total cost of supplying a unit of the resource by the i th producer would be comprised of total production costs and transportation costs. The production cost function can be written: $F_i + V_i(Q_i)$; and the transportation costs can be written: $\rho_i Q_i$, where ρ_i is the per unit transport cost. The total cost for i th producer of the biomass resource (C_i) would be the sum of the two cost components identified above.

$$C_i = F_i + V_i(Q_i) + \rho_i Q_i \qquad (22)$$

If it is assumed that we have perfectly competitive resource markets, the price faced by each producer would be the market price P. In order for the i th producer to maximize profits (π), the i th producer should supply additional units of the biomass resource until the marginal costs of production and unit transport costs are equal to the market price of the resource.

This is demonstrated mathematically below:

Profit Function

$$\pi = PQ - (F_i + V_i(Q_i) + \rho_i Q_i) \qquad (23)$$

First-order condition for profit maximization

$$\frac{d\pi}{dQ_i} = P - V_i'(Q_i) - \rho_i = 0 \qquad (24)$$

i.e. $P = V_i'(Q_i) + \rho_i$

Second-order condition for profit maximization

$$\frac{d^2\pi}{dQ^2} = -V_i''(Q_i) < 0 \qquad (25)$$

i.e. $V_i''(Q_i) > 0$

The implication of this analysis is that the supply curve for the i th producer would be the segment of the marginal cost curve plus the unit transport cost which exceeds the average variable cost curve plus unit transport cost. The optimal level of output for the i th producer would be attained when market price is identical with marginal costs plus unit transport cost. It is evident that in perfectly competitive resource markets with producers spatially separated, that producers with higher transportation costs would tend to supply less of the resource on the market than other producers which are located closer to the market (Henderson, 1958).

5. *Price Determination.* Figure 2 illustrates the determination of biomass fuel price in the short run under perfectly competitive resource market conditions. A perfectly competitive market for a resource or factor input has the following properties:

(1) The input is homogeneous,
(2) Large number of buyers and sellers,
(3) Buyers and sellers have perfect information,
(4) Free mobility of all buyers and sellers to exist or enter the market,
(5) Buyers are indistinguishable from the sellers standpoint.

The demand curves are denoted by D_i (i = 0,..,5) in order of increasing consumption responsiveness to price. The supply curve is denoted by S which is identical to the marginal cost curve up to output level Q_{MAX} beyond which no further output can be provided irrespective of price. At Q_{MAX} the relevant portion of the supply curve is shown by the vertical section of the supply curve (S). In this market situation, resource demand curve (D_0) would determine in conjunction with the marginal costs, a price P_0. The supply curve denoted in Figure 2 is perfectly elastic over the range of output up to Q_2, i.e., the cost of supplying or producing an additional unit is equal to the cost of providing the previous unit. As a consequence, demand can shift from D_0 to D_1, for example, without affecting market price.

Firms providing the resource will experience as an aggregate an increase in total revenues from P_0Q_0 to P_0Q_1. If the level of demand was D_2, the marginal costs of supplying the

Microeconomic Approaches to Biomass Fuel Pricing

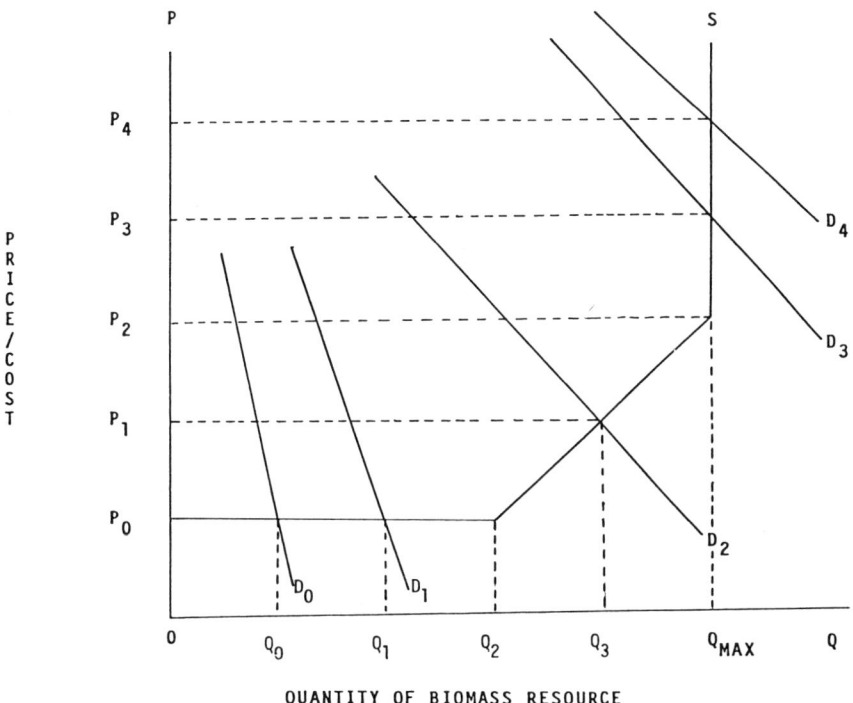

FIGURE 2. Determination of biomass fuel price short run competitive equilibrium.

resource quantity demanded would rise and equilibrium price would be determined at P_1. Finally, if the level of demand was so great as to exhaust supply, market price would ration the limited quantities of the resource available for purchase. This situation is illustrated in Figure 2 given demand curves D_3 and D_4. At D_3 the equilibrium price would be P_3 and supply would be Q_{MAX}. Given D_3, and a price P_2, the quantity demanded would exceed available output, under competitive conditions resource demands would bid up the price from P_2 to P_3 to obtain the available biomass resources.

C. Transfer Pricing

Transfer pricing pertains to the problem of product pricing between divisions or other segments of corporations or large business enterprises. The pricing problem presents itself because many industrial firms are characterized by diverse organizational structures and by multi-product production. For example, in the forest products industry, many large firms are vertically integrated-owning forest lands, sawmills, pulpmills, and they generate products such as wood chips, lumber, paper goods and biomass fuel resources. What price should be charged per unit for consumption of biomass fuel resources at the pulpmill if such resources were generated in the sawmill? In this section an economic analysis of transfer pricing is presented (Brigham and Pappas, 1972). The analysis considers resource products in competitive as well as imperfectly competitive markets.

1. Transfer Pricing: Perfect Competition. Figure 3 illustrates the determination of the appropriate transfer price on the assumption that the biomass resource is bought and sold in competitive markets. The biomass resource is used in the production of some final product (F). The demand for the final product is denoted by D_F and the associated marginal revenue by MR_F; the demand, marginal revenue, and marginal cost curves for the transferred biomass resource are denoted by D_T, MR_T, and MC_T respectively.

The marginal contribution to profits before deducting the transfer price is shown graphically by the curve denoted by $MR_F - MC_F$. The biomass resource price is determined in a perfectly competitive market (P_T). The enterprise can purchase all its requirements at that level. Thus, the demand curve D_T is perfectly elastic at price (P_T).

In order to maximize profits, marginal revenues should be equated to marginal costs for the final product as well as for the intermediate product. The profit maximizing points for the two divisions are shown in Figure 3. The division producing the final product (F) should purchase Q_F amount of the biomass resource at a price P_T. This occurs at point A where ($MR_F - MC_F$) equals P_T; profits would equal the area under the ($MR_F - MC_F$) curve above the line $P_T D_T$. It is evident that additional purchases of the resource up to point A result in positive net marginal contributions to profits but beyond point A additional purchases of the transferred product would reduce profits.

The division producing the biomass resource would maximize profits at point B where $MR_T = D_T = MC_T$ and output is Q_T. The division profits would be indicated by the area under the line $P_T D_T$ above the curve MC_T. The output Q_T would be supplied to the other division and that division should obtain the additional quantities needed ($Q_F - Q_T$) in the external competitive market. In the alternative situation where Q_T

Microeconomic Approaches to Biomass Fuel Pricing

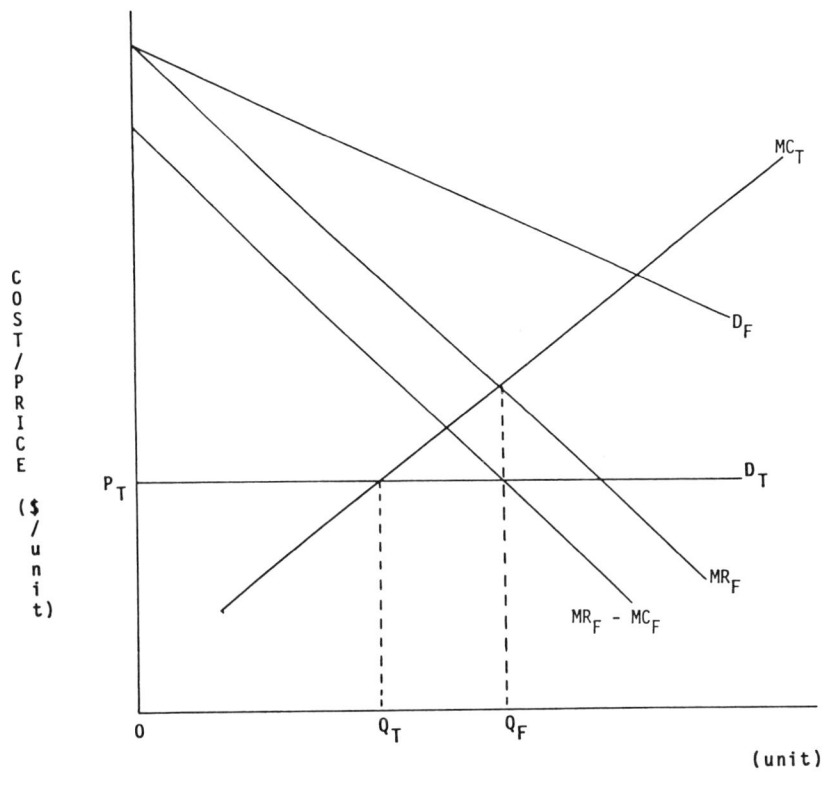

FIGURE 3. Transfer price determination with biomass resource price determined in competitive markets.

exceeds Q_F, the intermediate product producing division supplies the amount needed by the other division (Q_F) and sells the extra quantitites ($Q_T - Q_F$) on the open market.

The basic conclusion of this analysis is that an intermediate good transferred between divisions should have a transfer price equal to the market price providing the good is bought and sold in a perfectly competitive market.

2. Transfer Pricing: Imperfect Competition. The determination of the appropriate transfer price for an intermediate good sold in imperfectly competitive markets and transferred between divisions is illustrated in Figure 4. Panel (a) illustrates the demand and net marginal revenue curves for the

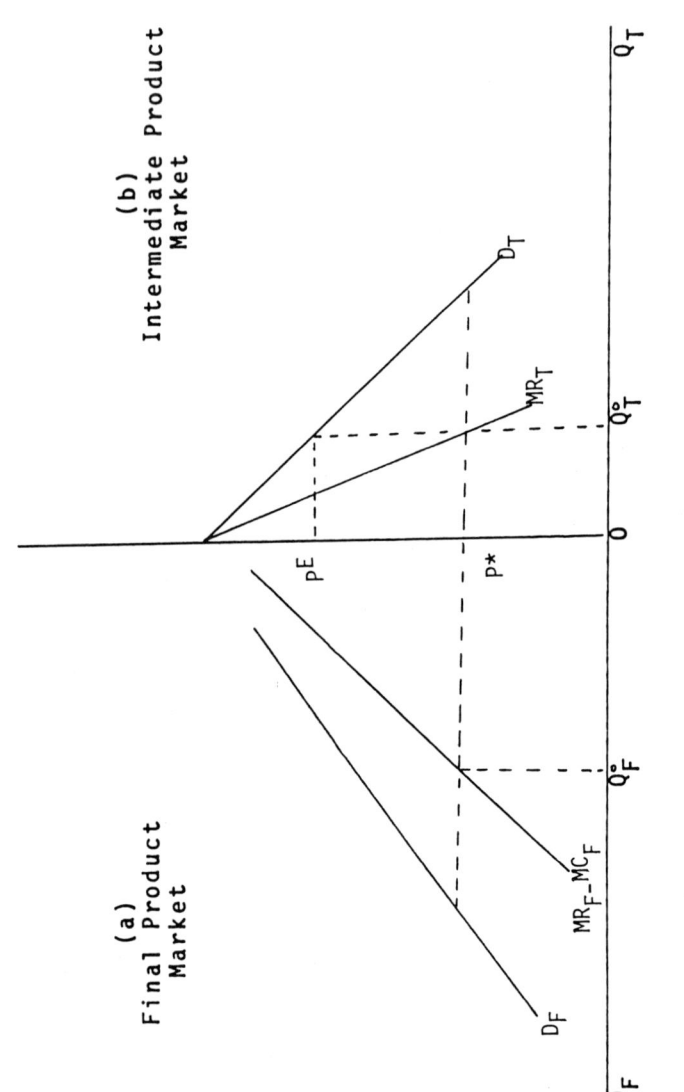

FIGURE 4. *Transfer price determination with biomass resource price determined in imperfectly competitive markets.*

final product of the company; panel (b) shows the demand and marginal revenue curves for the transferred resource. If the transferred resource is sold in an imperfect market, its demand curve would be downward sloping and the marginal curve would lie below the demand curve.

The price P* represents: (1) the price at which $MR_T = MR_F - MC_F$ and (2) the price where marginal cost of transferred resource (MC_T) equals the aggregate net marginal revenue for the company. The aggregate net marginal revenue curve is the horizontal summation of ($MR_F - MC_F$) and MR_T.

From the company's standpoint, it would tend to optimize profits at that level of output where the marginal costs of supplying the transferred resource equals the net marginal revenue associated with the resource. The distribution of the total output of the transferred resource Q_T, would be between the internal or divisional market and the external market. Because of imperfect nature of the market, the company would act as a discriminating monopolist selling a portion of the output of the resource internally at a price P* which equals MC_T and the remaining output in the outside market at a higher price P^E.

3. *Transfer Pricing: Nonmarket Case.* In this section, transfer pricing in the absence of an external market is examined. Market price does not provide a basis to value resource transfers within the company. How does one determine then what the price should be that a company would have one division pay another supplying division of the same company? The basic answer is that the transfer price should be based on the marginal costs of producing the intermediate product. The fundamental economic arguments behind this rule are explored in the following discussion and illustrated in Figure 5.

Assume that the company produces a final product F which is sold in the product market. The company maximizes its profits by supplying F until the marginal revenue associated with an increment of output equals the marginal costs of supplying the product. Figure 5 illustrates the case taken in account the effect of the transferred resource on profits. The optimal output of the intermediate product or resource input occurs at Q* where the net marginal revenue is equal to the marginal costs of producing the transfer resource. The price P* that the supplying division should charge the other division using the resource is determined by the marginal costs of producing the intermediate resource. To the left of point A, the net marginal revenue curve ($MR_F - MC_F$) exceeds the marginal cost of the resource (MC_T) which implies that supplying more of the biomass resource in the production of product F would increase profits at the margin. However, if

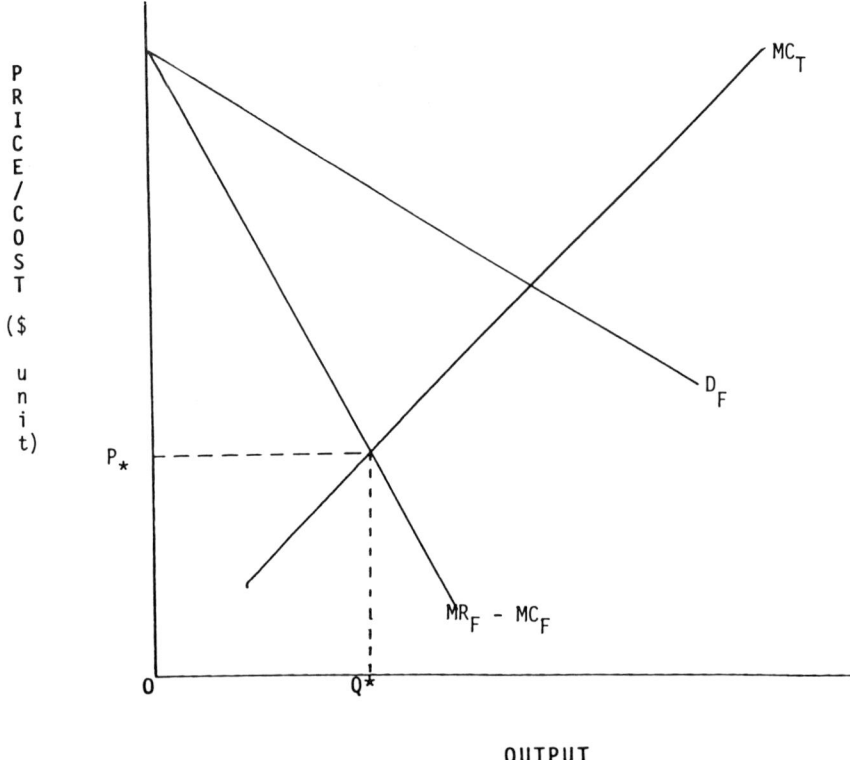

FIGURE 5. *Transfer price determination without external markets for biomass resource.*

the resource use is expended past Q_*, profits would tend to diminish because increments to cost (MC_T) exceed incremental revenue ($MR_F - MC_F$). Therefore, only at point A where $MC_T = MR_F - MC_F$ would overall company profits be maximized. In conclusion, the appropriate transfer price would be determined by the marginal costs of the intermediate resource.

This section has examined the problem of transfer pricing in two broad contexts: case of an outside market for the intermediate resource; and the case of no external market for the resource input. In the case of an outside competitive market, resource transfers within a company should be based on market prices. If the resource input is not traded in an outside market or there is an imperfect resource market, transfer pricing should be based on marginal cost.

IV. APPLICATION OF MICROECONOMIC CONCEPTS

This section presents three areas where the theoretical concepts are applied to biomass resource pricing problems. The first case provides an example of marginal cost pricing of wood fuel employed in a cogeneration facility. The second example is concerned with the replacement of fossil fuels with wood fuel in existing industrial technologies. In this example, a value to the biomass fuel is imputed based on alternative costs concepts. Both of these case examples were taken from actual studies using industry data.

Pricing of resources in a multi-division firm is addressed in the third example. The linear programming approach is presented and offered as a method to impute values to productive resources which would determine appropriate prices for the intra-firm resource transfers. This example highlights the use of linear economic models to analyze the pricing problem and the use of mathematical programming techniques to provide solutions.

A. *A Supply and Demand Analysis for Wood Fuel*

This biomass fuel pricing example is taken from an electricity feasibility study of a proposed cogeneration facility in Hulett, Wyoming (Ebasco, 1982). A significant feature of this study is the development of an incremental cost curve for wood waste fuel.

The technical aspects of the proposed facility are presented first to establish the firm's resource demand requirements and then the local supply characteristics of the wood-waste market are discussed. After these topics are covered, the marginal or incremental cost of wood fuel is constructed for the Hulett market. Given the demand conditions existing in the Hulett wood fuel market, the price that the cogenerator would have to pay for wood-waste fuel is determined.

1. Proposed Cogeneration Facility. The largest sawmill in the community of Hulett, Wyoming, has the potential to be the site of a cogeneration facility. This potential is based on the available supply of wood residues in the area and process steam load present at the mill. Three electric power generation alternatives have been deemed feasible: (1) A

527 kW back pressure turbine based cogeneration system, (2) a 2300 kW extraction turbine based cogeneration system, and (3) a 9200 kW condensing turbine based power plant.

Each of the power options listed above are designed to produce electricity taking into consideration the process steam requirements at the mill (average - 13,000 lbs/hour) and a total fuels availability of 57,000 oven dry (O.D.) tons/year. The consumption of wood fuels changes commensurately with the electrical output associated with running the cogeneration schemes. Natural gas has been proposed as the secondary fuel for the cogeneration facility.

Table II provides summary statistics for the three cogeneration options. It is anticipated that Options 1 and 2 will be operated 8,000 hours during the year which represents a capacity factor of about 13,900 green tons/year or 8,620 O.D. tons/year. Option 3 would consume 47,300 O.D. tons/year and would be operated about 7,000 hours each year.

Table III provides annual estimates of wood residue quantities that are available at various distances from Hulett. Options 1, 2 and 3 would account for 15, 38 and 83% respectively of the total annual wood waste in the Hulett, Wyoming, area; i.e., 57,000 O.D. tons/year.

2. *Development of Marginal Cost Curve.* At the large sawmill, bark and sawdust residues are produced which can be sold locally to various users. The quantity of these byproducts of sawing lumber is about 12,000 tons per year; and the price that the mill can obtain for these resources is about $5/ton. The price is expressed F.O.B. at the sawmill. There are about 10,000 tons of wood chips available at the sawmill which could be sold locally at $7 to $8 per ton (F.O.B.). Because these sawmill byproducts have positive opportunity costs, they should not be treated as byproducts with nil value. The opportunity cost is measured by the market price that these wood products can command on the local markets.

To obtain wood fuels in excess of 26,000 tons per year, wood products must be purchased from sources which lie farther and farther away from the proposed site of the cogeneration facility. The quantities available from spatially separated producers outside Hulett are shown in Table III. Although these quantities are available, transportation costs rise in proportion to the distance between the source and the cogeneration facility. The transportation costs are expected to add an average of about 20 cents/ton/mile to the production costs of the resource. Figure 6 illustrates the incremental or marginal cost curve of fuel delivered to the sawmill.

TABLE II. Comparative Summary of Three Cogeneration Options

Key parameters	Options		
	1	2	3
Electrical capacity (kW)	527	2,300	9,200
Heat rate (Btu/kWh)	5,200	16,000	14,800
Total fuel consumption rate (lbs/h)	3,478	8,785	28,900
Installed cost ($/kW)[a]	1,600	2,300	2,100

[a] Cost chargeable to electric power only.

TABLE III. Estimated Quantities of Wood Residues Available in Hulett, Wyoming, Area

Radius from Hulett (Miles)	Aggregate total (O.D. tons)
0	57,000
less than 55	74,000
less than 80	90,000
less than 150	136,000

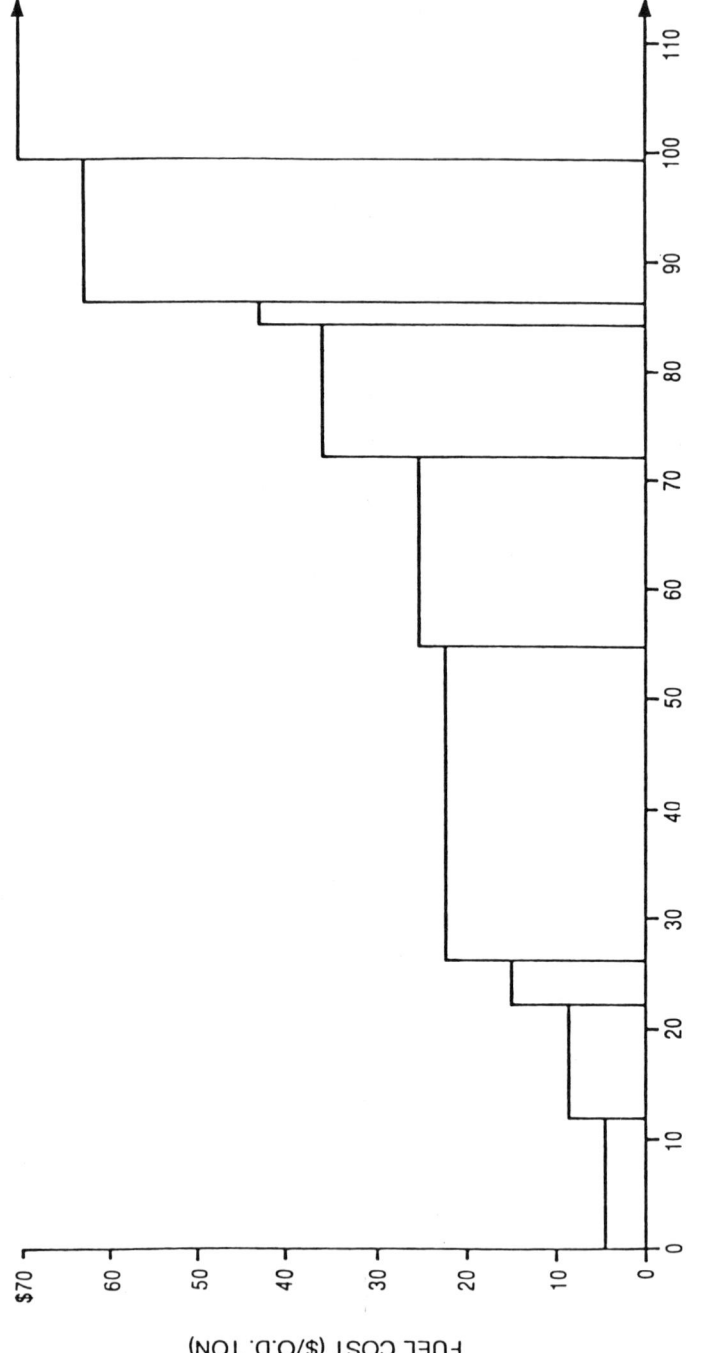

FIGURE 6. Marginal cost curve of fuel delivered to sawmill.

3. *Development of Demand Curve and Price of Fuel.* To determine the purchase cost of wood fuel for the cogeneration facility, the demand requirements at the facility must be identified. After the capacity (kW) of the cogeneration facility has been determined as well as its operating regime, the annual fuel requirements can be estimated. Although the input requirements are known, the demand curve would not be totally inelastic due to interfuel substitution possibilities and the effects of stockpiling. Thus the demand curve is expected to be downward sloping denoting some degree of own price elasticity.

The intersection of the demand curve with the supply curve would determine the marginal cost price that the cogeneration facility's owner is willing to pay to obtain the desired quantities of wood fuel. The marginal cost curve reflects the amount of wood fuels that resource owners are willing to provide at the given prices. If the level of demand for wood fuel were low, shavings at the sawmill might be the marginal fuel and the appropriate price or purchase cost of wood fuel would be $15/ton. However, if the level of demand was greater, the incremental fuels would be obtained from nonlocal sources and the marginal cost price would rise reflecting the higher costs associated with transportation of the wood fuel. The fuel price for Options 1, 2 and 3 would be $5/ton, $8/ton, and $22/ton respectively based on the annual fuel requirements and the marginal cost curve in Figure 6.

B. *An Alternative Cost Approach to Value Wood Fuel*

In many manufacturing or industrial plants, various technologies are employed in the production process which can use alternative fuels. Because fuels may be substituted for one another in production, their relative prices are important in the selection of the primary fuel. Changes in relative fuel prices may be significant enough that interfuel substitution occurs even if retrofitting of existing technology is necessary. It follows that the value of one fuel can be approximated by the price of a close substitute fuel. For example, natural gas has been valued at the price of residual fuel oil or distillate oil in certain market areas of the United States in recent years.

The alternative cost approach to value wood fuels is demonstrated in the context of industrial wood fuel using technologies in the following case example. The case example has been taken from a 1980 study by the Congress of the United States (Office of Technology Assessment, 1980).

1. *Economic Analysis of Industrial Wood Fuel Using Technologies*. The economic analysis consists of imputing values to wood fuel based upon the costs of alternative technologies and fuels. In the case of direct combustion, imported oil is the substitute or alternative to wood fuel. With cogeneration, the alternative chosen for analysis is raising process steam with wood and purchasing electricity. In the case of generating electricity, value is placed on the power benefits.

The value of wood fuel is approximated by the difference in total annualized capital and O & M between the wood fired system and the alternative fuel system relative to total wood fuel usage. The specific formulae employed in the calculations are listed below:

a. Calculation formulae

Formula for combustion

$$WV_S = \frac{(C_c + C_{o\&m} + C_f)_p - (C_c + C_{o\&m})_w}{WC_S} \qquad (26)$$

WV_S = Value of wood fuel used to raise steam

C_c = Annual capital cost

$C_{o\&m}$ = Annual operating and maintenance cost

C_f = Annual cost of fuel

p = Given sized petroleum fired system

w = Comparably sized wood fired system

WC_S = Annual tons of wood fuel consumed (green wt, $ M specified) to raise steam

Assumptions: Oil as numeraire – either as the alternative or as a way of pricing coal. Wood not priced.

Formulae for cogeneration
Formulae for cogeneration increment

$$WV_e = \frac{(E_p \times E_v) - \Delta(C_c + C_{o\&m})}{WC_{cg}} \qquad (27)$$

WV_e = Wood value for that increment of fuel required for cogeneration

E_p = Electricity produced annually by cogenerative system

E_v = Value of electricity produced, as replacement for energy purchased or as sold to the power grid

ΔC_c = Annual additional capital cost required for cogeneration

$\Delta C_{o\&m}$ = Annual additional operating and maintenance cost required for cogeneration

WC_{cg} = Additional wood fuel increment consumed annually to support cogeneration.

Formula for all fuel burned in a cogeneration facility

$$WV_t = \frac{[(C_c + C_{o\&m} + C_f)_p - (C_c + C_{o\&m})_w] + [(E_p \times E_v) - (C_c + C_{o\&m})]}{WC_s + WC_{cg}} \quad (28)$$

Formula employed for electricity generation systems

$$WV = \frac{[P_a \times P_v] - [C_v + C_{o\&m}]}{WC_a} \quad (29)$$

Where

WV = Wood value ($/ton)

P_a = Annual product production (e.g., gal, KWH)

P_v = Produce value ($/gal, $/KWH)

C_c = Annual capital cost, wood system

$C_{o\&m}$ = Annual operating and maintenance cost, wood system

WC_a = Wood consumed annually (green tons, moisture specified)

b. *Capital cost assumptions and estimates.* Table IV summarizes annual capital cost figures for a 25-year plant life under competitive industry and utility industry conditions. The item covered in the table are expressed as a percent of initial plant investment.

2. *Analysis of Process Steam Generation.* Table V presents the basic cost and savings data for a 120,000 lb/hour boiler assuming the generation of process steam. These data comprise the base case for direct combustion analysis.

Table VI converts that data into imputed fuel values for wood on both a green ton and bone dry ton (BDT) according to formula 26. This table examines the effect of moisture content on the imputed value of wood.

3. *Analysis of Cogeneration.* Table VII adds the capital cost of the high pressure/high temperature (HP/HT) values, the feedwater pretreatment system, and the turbogenerator required for cogeneration to the basic 120,000 lb/hour combustion system. Then, cogeneration costs are compared to the

TABLE IV. *Annual Capital Cost Item as a Percentage of Initial Plant Investment (25-Year Plant Life)*[a]

Item	Type of industry	
	Competitive	Utility
Depreciation	4.00	4.00
Debt	0.00[b]	4.50[c]
Equity	15.00[d]	3.75[e]
Income tax (@ 48%)	14.40	3.60
Total	33.40	15.85

[a] For each case, the annual capital costs are a critical factor. Competitive industry conditions are assumed for each alternative case except for the electricity case. The economic analysis of wood value assumes moisture content at 50% level.

[b] 0%.
[c] 75% at 12% IRR
[d] 100% at 30%/annum
[e] 25% at 30%/annum

TABLE V. Imputed Value of Wood Fuel as a Function of Oil Saved in a Wood Fired Boiler (1978$)[a]

Cost category	Wood system ($)	Oil System ($)
Capital cost	6,768,000	1,647,000
Site preparation	106,000	93,000
Fuel handling	939,000	100,000
Boiler related	3,692,000	943,000
Spare parts	150,000	50,000
Direct cost	4,887,000	1,186,000
Engineering	733,000	178,000
Contingencies	843,000	205,000
Working capital	305,000	78,000
Annual capital cost @ 33.4%	2,261,000	550,000
Annual operating and related costs	825,000	4,026,000
Labor and utilities	249,600	124,800
Maintenance	406,100	98,800
Local taxes and insurance	169,200	41,200
Fuel (oil @ 14.23/bbl) (264,300 bbls)	–	3,761,000
Total annual cost	3,086,000	4,575,000
Annual savings attributable to fuel	1,490,000	

[a] Assumes 120,000 lb/hr, 150 psig. process steam exhausted from backpressure turbine.

Source: Tillman, 1979.

TABLE VI. Influence of Moisture on Imputed Fuel Value: Combustion Case (120,000 lb/hr steam) ($000).

Cost item	Fuel moisture content			
	50% M	25% M	10% M	Oil
Installed capital	6,800	6,800	6,800	1,650
Annual capital	2,260	2,260	2,260	550
Annual O & M	825	825	825	265
Fuel cost (oil-1978)				3,760
Total annual cost	3,085	3,085	3,085	4,575
Annual savings	1,490	1,490	1,490	
Annual tons consumed (tons)	215,000	122,000	92,000	N/A
Value of wood fuel ($/ton)	6.95	12.20	16.20	
Value of wood ($/BDT)	13.90	16.30	18.00	

alternative of purchasing power at 28 mills/KWH and 35 mills/KWH, the 1978 national average for industrial power, and marginal cost for generating new power, respectively.

Table VIII converts base case data into imputed values as a function of fuel moisture content and electricity price. It makes the assumption in the case of generating replacement power that there would be no standby power cost. If relevant, values for such charges can readily be added to annual operating costs.

4. *Electricity Generation*. Electricity generation economics are based on the lower capital costs shown in Table IV for the utility industry than the higher capital costs in private industry.

Table IX presents base case data for the 55 MW electric power plant, the limit of practical scale for a total biomass energy system. It should be noted that each KWH costs 27.7 mills before fuel cost is considered. Table X imputes values to the wood fuel based upon formula (29) using marginal costs of power of 35 and 50 mills/KWH.

TABLE VII. Imputed Value of Wood Fuel Used in Cogeneration of In-Plant Needs as a Function of Electricity Price[a] (1978$)

Cost category	Basic combustion system costs ($)	Additional cost for cogeneration ($)
Total capital cost	6,768,000	1,552,000
Annual capital cost @ 33.4%	2,261,000	520,000
Annual operating costs	825,000	209,000
Labor and utilities	249,600	50,400
Maintenance	406,100	120,000
Local taxes and insurance	169,200	38,800
Electricity purchases (40 × 10^6 KWH @ 28 mills/KWH)	1,120,000	
Electricity purchases (40 × 10^6 KWH @ 35 mills/KWH)	1,400,000	
Incremental annual cost		729,000
Total annual cost (28 mill case)	4,206,000	3,815,000[b]
Total annual cost (35 mill case)	4,486,000	
Annual savings by cogeneration (on 28 mill power case)		391,000
Annual savings by cogeneration (35 mill power case)		671,000

[a] Base case information: Assumes 5 MW system, 120,000 lb/hr steam.

[b] $4,206,000 − $1,120,000 + $729,000.

TABLE VIII. *Cogeneration Economics as Influenced by Moisture Content and Electricity Price (1978$)*

Cost item	Fuel moisture content					
	50% M		25% M		10% M	
	28 Mill power	35 Mill power	28 Mill power	35 Mill power	28 Mill power	35 Mill power
Annual savings ($)	391,000	671,000	391,000	671,000	391,000	671,000
Incremental annual wood consumed (tons)	29,200	29,200	16,600	16,600	12,500	12,500
Value of congeneration wood incr. ($/ton)	13.40	23.00	23.55	40.40	31.30	53.70
Total wood consumed (tons)	244,200	244,200	138,600	138,600	105,000	105,000
Value of all wood ($/tons)	7.70	8.85	13.50	15.60	17.90	20.60
Value of all wood ($/BDT)	14.40	17.70	18.05	20.80	19.90	22.85

TABLE IX. The Influence of Electricity Price and Moisture Content on Imputed Wood Value in Power Generation (1978$)

Item	Fuel moisture content		
	50%	25%	10%
Total annual cost ($000)	12,200	12,200	12,200
Total annual wood consumed (10^3 tons)	$1,035^a$	520^b	375^c
Annual product value @ 35 mills/KWH ($000)	15,400	15,400	15,400
Wood^d value ($/ton) @ 35 mills/KWH	3.10	6.15	8.55
Wood value ($/BDT) @ 35 mills/KWH	6.20	8.20	9.50
Annual product value @ 50 mills/KWH ($000)	22,000	22,000	22,000
Wood^d value ($/ton) @ 50 mills/KWH	9.45	15.00	20.80
Wood value ($/BDT) @ 50 mills/KWH	18.90	25.10	29.05

^{a}Heat rate = 20,000 Btu/KWH.
^{b}Heat rate = 15,000 Btu/KWH.
^{c}Heat rate = 13,000 BTU/KWH.
^{d}Imputed wood value = $\dfrac{\text{annual product value} - \text{annual cost}}{\text{annual wood consumed}}$

TABLE X. Base Case Information for the Imputed Value of Wood Based on Electricity Production[a] (1978$)

Cost item	Base case cost ($000)
Direct capital cost	37,800
Engineering	5,700
Contingencies and commissions	6,500
Working capital	2,500
Total installed cost	52,500
Annual capital (CRF @ 15%)	8,300
Annual operating costs	3,900
Total annual cost	12,200
Total electricity produced (8000 hours)	440,000,000 KW
Cost without fuel	27.7 mills/KWH

[a] Assumes 50 MW.

C. Transfer Pricing: A Linear Programming Approach

The fundamental theory of transfer pricing was presented in section III. In practice, applying the theory can be very difficult due to the number of products and variable resource inputs, resource and output constraints, and the degree of decentralization decision making permitted within the organization of the firm. Linear programming models have been found to be powerful tools in addressing problems concerned with the allocation of scarce resource amongst competing users in an optimal manner.

1. *Linear Programming Model.* A linear programming problem is that of finding nonnegative numbers $X_1, \ldots, X_m$ which either maximize or minimize a given linear function (Z):

$$Z = \sum_{i=1}^{m} c_i X_i \tag{30}$$

where the numbers X are required to satisfy a set of linear inequalities

$$\sum_{i=1}^{m} A_{ij} X_i \le B_j \qquad j = 1, \ldots, n \qquad (31)$$

A significant feature of linear programming is that a minimum problem corresponds to the maximum problem. The standard minimum problem would be stated as finding nonnegative numbers $Y_1, \ldots, Y_n$ which minimize

$$\sum_{j=1}^{n} B_j Y_j \qquad (32)$$

subject to the set of inequalities

$$\sum_{j=1}^{n} A_{ij} Y_j \ge C_{ij} \qquad i = 1, \ldots, m \qquad (33)$$

This feature is the fundamental duality theorem which can be stated as follows: If a standard maximum or minimum problem and its dual are both feasible then they both have optimal solutions and both have the same value. If either problem is not feasible then neither has an optimal solution (Gale, 1960).

2. *Economic Interpretation of Dual Problem.* Before interpreting the dual problem, let us assume that the primal problem is one of profit maximization. In this case the variables and parameters of the primal problem may be viewed as follows:

Quantity	Interpretation
X_i	Level of activity i (i = 1, ..., m)
C_i	Unit profit from activity i
Z	Total profit from all activities
B_j	Amount of resource j available (j = 1, ..., n)
A_{ij}	Amount of resource j consumed by each unit of activity i.

In the dual problem, Y_j represents the marginal value or contribution of resource J (j = 1, ..., n) to profits. The values of B_j represent potential or anticipated allocation of resources to the set of activities. The dual variables or shadow prices of the resources are used to evaluate whether the existing resource mix should be reallocated. The shadow price Y_j^* for resource j represents the maximum unit price one would offer to increase the use of resource j. If the shadow price exceeds actual unit cost, the resource is contributing more to profit than cost at the margin and thus resource should be increased in amount until the inequality vanishes. Whereas the optimization problem for our primal problem is maximization of profits, the objective of the dual problem is the minimization of the implicit value or cost of consuming all resources in various employments or activities.

The shadow prices are also called accounting prices since they have been determined by imputing value to the resources based on their contribution to the total profits of the firm (at least in the model formulation assumed for discussion purposes). Because these shadow or accounting prices can be calculated by solving the linear programming problem (if a solution exists), linear programming methods can be used for imputing resource values or prices in many applied economic problems such as transfer prices for intra-firm resource transfers. This is important when market prices for the resources are unavailable or existing market prices are distortionary.

Duality theory along with its theorems are important because they not only increase our understanding of linear programming but also aid in the solution of programming problems and have an economic interpretation corresponding to marginal economic analysis (Baumol, 1972). A more rigorous treatment of linear economic models and duality theory can be found in the work of Gale (1960). The work of W. J. Baumol provides an excellent discussion of the economic interpretation of the dual problem (Baumol, 1972). There are many sources devoted to linear programming and related techniques. However, the operations research text by Hillier and Lieberman (1974) is recommended not only for its coverage of mathematical programming but also its analysis of multidivisional problems and the principle of decomposition. It is beyond the scope of this chapter to present the theory of linear programming and techniques for solving such models. As a consequence, the researcher is referred to the basic texts identified above for in-depth treatment of the subject matter.

V. CONCLUSIONS

Microeconomic theory offers several powerful approaches for appropriate pricing of biomass resources. Though biomass resources are indeed multifaceted and their use as a fuel source is less established the standard principles of supply and demand in the market place still apply. Biomass fuels may be priced at their opportunity costs; that is, the price that such resources would command in other employment or uses. When resource markets are imperfect or nonexistent, the marginal costs of production can be used to evaluate biomass fuels.

The alternative cost concept has been shown to be useful in the process of valuing biomass resources used in the production of such products as process steam and electricity. The benefits of biomass generated power for example was estimated by the cost of the next best alternative fuel technology. By computing the differential in costs between the biomass and alternative fuel technology, a value can be imputed to the biomass fuel.

Finally, "shadow prices" or accounting prices can be calculated by linear programming techniques and then used to impute prices to intra-firm resource transfers. These shadow prices correspond closely with marginal cost theory and provide yet another approach to set biomass resource prices.

The literature suggests that very limited application of microeconomic theory has been undertaken to date. As a result, the pricing of biomass resources or fuels has been done in many cases in an arbitrary manner. If this chapter serves to highlight the usefulness of microeconomic concepts and encourages further application of the theory for biomass resource pricing, it would have achieved its purpose.

REFERENCES

Baumol, W. J. 1972. "Economic Theory and Operations Analysis." Prentice Hall, Englewood Cliffs, New Jersey.

Benemann, J. R. 1980. Biomass Energy Economics. *The Energy Journal,* Vol. 1, No. 1, pp 107-131.

Congress of the United States, Office of Technology Assessment. 1980. "Energy from Biological Processes," Vol. III, Appendixes Part A: Energy from Wood. Washington, D.C.

Brigham, E. G., and Pappas, J. L. 1972. "Managerial Economics." Dryden Press, Hinsdale, Illinois.

Deaton, A., and Muellbauer, J. 1980. "Economic and Consumer Behavior." Cambridge University Press, London.

Ebasco Services, Inc. 1982. "Hulett, Wyoming, Woodwaste Cogeneration Project." Bellevue, Washington.

Ferguson, C. E. 1972. "Microeconomic Theory." Richard D. Irwin, Inc., Homewood, Illinois.

Gale, D. 1960. "The Theory of Linear Economic Models." McGraw Hill Co., New York.

Henderson, J. M., and Quandt, R. E. 1958. "Microeconomic Theory." McGraw-Hill Co., New York.

Hillier, F. S., and Lieberman, G. J. 1974. "Operations Research." Holden-Day, Inc., San Francisco.

Slesser, M. and Lewis, C. 1979. "Biological Energy Resources." E & F N. Spon Ltd., London.

Tillman, D. A. 1979. The Economic Values of Wood Residues as Fuel. In "Progress in Biomass Conversion," Vol. 1. Academic Press, New York.

FUEL CHARACTERISTICS OF SELECTED SPECIES OF BEACHED LOGS IN SOUTHEASTERN ALASKA

W. Ramsay Smith

University of Washington
Seattle, Washington

Richard O. Woodfin, Jr.

Pacific Northwest Forest and Range Experiment Station
USDA Forest Service
Portland, Oregon

I.	INTRODUCTION	204
II.	PROBLEMS OF FUEL CONTAMINATED WITH SALT	204
III.	SALT CONTENT OF BEACHED LOGS	205
IV.	HIGHER HEATING VALUES OF BEACHED LOGS	210
V.	PROXIMATE ANALYSIS OF BEACHED LOGS	210
VI.	EFFECTS ON INDUSTRIAL OPERATIONS OF USING BEACHED LOGS	213
VII.	CONCLUSIONS	215

I. INTRODUCTION

Timber from the Tongass National Forest in the Alexander Archipelago of southeastern Alaska is rafted through various straits to sawmills in the area. Harsh and sudden changes in weather conditions have caused the log rafts to break up leaving them scattered and freely floating in the waters. This material eventually washes onto beaches. Currently the accumulation of these logs involves billions of board feet.

The large quantities of material involved have prompted several studies by the USDA Forest Service, Pacific Northwest Forest and Range Experiment Station, to determine its utilization potential. One of these studies was done in cooperation with the College of Forest Resources, University of Washington, to determine problems that may be encountered if this material is burned in industrial wood-fired furnaces.

Potential problems in combustion could result because of inorganic contaminants picked up during the exposure of these logs to saltwater and beach environs. The objective of the study was, therefore, to determine the sodium and other inorganic contents of sitka spruce and western hemlock, taking into account the log diameter and radial position. In addition, other fuel characteristics (i.e., higher heating values and proximate analysis) were determined for comparative purposes.

II. PROBLEMS OF FUEL CONTAMINATED WITH SALT

Salt contained in fuel has been found to cause problems in boilers whether the fuel is wood or oil. Leman (1975) discussed some of the problems that occur when bark from logs that have been transported and stored in saltwater is burned. Salt in hog fuel corrodes metal components in conveyor systems. Such components then have to be replaced with plastic components. He also found that between 65% and 85% of the salt contained in the bark passed out the boiler stack as particulate emissions, increasing plume opacity beyond acceptable levels. These particulates are submicron in size, ranging from approximately 0.4 μm to 1.0 μm, and they are very difficult to capture. Because these salt particulates have a high potential for scattering light, they produce high opacity readings.

Various types of collection equipment have been examined to remove the salt particulate (Brady and Jenkins, 1980). Wet and dry scrubbers can remove sufficient quantities of salt, but not without associated problems. Wet scrubbers have a difficulty with disposal of the effluent. Dry scrubbers

require a high pressure drop to be effective, and therefore
such devices become energy intensive. Baghouses are the most
effective, but corrosion can result if condensation occurs.
The baghouse, therefore, must be well insulated with a minimum
of 7.5-cm fiber insulation and be supplied with external heat
when not in use.

The salt remaining in the fuel, which lowers overall
boiler efficiency and disrupts uniformity of air flow through
the fuel bed, is deposited on grates, boiler tube surfaces,
and breachings. This salt increases the ash content of the
fuel and can lower its fusion point from approximately 1300°C
to 927°C, causing unwanted slagging. Once in the ash stream,
salt also causes corrosion in ash handling equipment.

High temperature and low temperature corrosion caused by
presence of sodium chloride in fuel oil was discussed by
Goldberg and Bennett (1980) and Babcock and Wilcox (1978).
Concentrations of 70 parts per million (ppm) of sodium were
found to corrode waterwall tubes, convection surfaces, supports, and spacers. The sodium chloride component decomposes
in the combustion chamber and reacts with oxygen and hydrogen
to form Na_2O and HCL. If sulfur is present, the sodium will
react further to form Na_2SO_4.

One additional area where sodium is destructive in oil-
fired systems is with high aluminum content refractories with
relatively porous structure that allows penetration of sodium
compounds. These compounds react with the aluminum in the
bricks to form sodium metaaluminate $(NaAL)_2$. It has a high
melting point (1800°C) and, therefore, will not slag under
these conditions but will expand because of its thermal expansion properties. The results may be the cracking and breaking
of the refractory.

III. SALT CONTENT OF BEACHED LOGS

Logs ending up on beaches have lost all bark so the wood
has been directly exposed to saltwater. Salt penetration depends on duration in the water, duration on the beach, and log
diameter. To investigate these variables, disks were cut from
beached logs of various diameters. The inorganics and fuel
characteristics were then determined at three points on the
log. These points were located at the surface, at one half
the log radius, and at the log centers. All disks were taken
more than 3 feet from the ends to minimize sorption effects
at the ends.

Determination of the time that the logs were in the water
and on the beach was not possible. Originally it was thought
this information could be approximated by log color and

location on the beach. During sampling, however, it was found that differentiation by these variables was not feasible. All logs were severely grayed from weathering, and were washed together in piles.

Major species sampled were western hemlock (Tsuga heterophylla (Ref.) Sarg.), sitka spruce (Picea sitchensis (Borg.) Carr.) and Pacific silver fir (Abies amabilis Dougl. at Forbes). Log diameters for western hemlock ranged from 14 to 74 cm, for sitka spruce from 18 to 69 cm, and for Pacific silver fir from 18 to 38 cm.

Analysis of the samples for sodium, iron, potassium, calcium, and magnesium were made using an atomic absorption spectrophotometer. The results are given in Table I.

The results for the beached logs were compared with inorganic content of trees located within a mile of the Oregon coast (unpublished work by G. C. Grier, 1982. A Tsuga heterophylla-Picea sitchensis ecosystem of coastal Oregon: Production and nutrient cycling). As shown in Table I, the sodium content of all three species is much higher for the beached logs than that for the control; 16 to 23 times greater in western hemlock, 14 to 22 times greater in sitka spruce, and 18 to 23 times greater in Pacific silver fir.

Because of the magnitude of the standard deviations, no significant differences were found with respect to radial position. This is shown in Figure 1. It was first thought to indicate an influence by the varying log diameters. An analysis was made of sodium content by log diameter and radial position for the three major species. Again, no significant differences were found with respect to radial position which would indicate that great variance is attributable to the log's history; i.e., duration in the water and on the beach. Unfortunately, as discussed previously, neither period of time could be determined for these logs.

Another cause of variation could be the location of the sample within the log. It was not noted at the time of sampling whether the sample wedge was obtained from the portion of the log facing the sky or facing the beach. Surface leaching of the salt by rain may be greater for the less protected top side, thereby tending to reverse the profile.

Both calcium and magnesium contents were also found to be higher in the beached material than in the control; this would be expected with an increase in saltwater content. Iron was not found to be significantly different.

Potassium was high in the western hemlock control material and thought to be due to the particular site characteristics. For more inland material, Edmonds[1] found average potassium

---

[1]*Personal communication of unpublished results. R. Edmonds 1982. College of Forest Resources, Univ. of Wash., Seattle.*

TABLE I. Inorganic Content (ppm)

Species	n	Na $\bar{X}$	Na σ	Fe $\bar{X}$	Fe σ	K $\bar{X}$	K σ	Ca $\bar{X}$	Ca σ	Mg $\bar{X}$	Mg σ
Western Hemlock											
Control	12	100	7[a]			800	7[a]	300	7[a]	100	7[a]
Surface	57	2310	1966	31	55	244	238	549	191	517	243
1/2 radius	54	1600	1705	21	66	353	459	592	254	238	300
Pith	56	1824	1776	6	42	470	474	770	231	272	306
$\bar{X}$		1917		19		355		637		345	
Spruce											
Control	12	100	7[a]			100	7[a]	200	7[a]	100	7[a]
Surface	17	2238	1758	tr	—	149	83	456	218	482	210
1/2 radius	17	1403	803	25	132	126	53	362	123	102	76
Pith	17	1337	913	tr	—	153	85	472	214	140	165
$\bar{X}$		1659		—		143		430		241	
True Firs											
Control	12	100	7[a]			100	7[a]	200	7[a]	100	7[a]
Surface	13	2151	1171	5	57	208	159	565	287	595	259
1/2 radius	13	1848	1188	tr	—	239	127	741	431	239	166
Pith	12	2321	1412	tr	—	434	220	975	309	366	291
$\bar{X}$		2101		—		290		755		401	

[a] Coefficient of variation in percent. Values obtained for control from unpublished results. (Grier, 1982, unpublished).

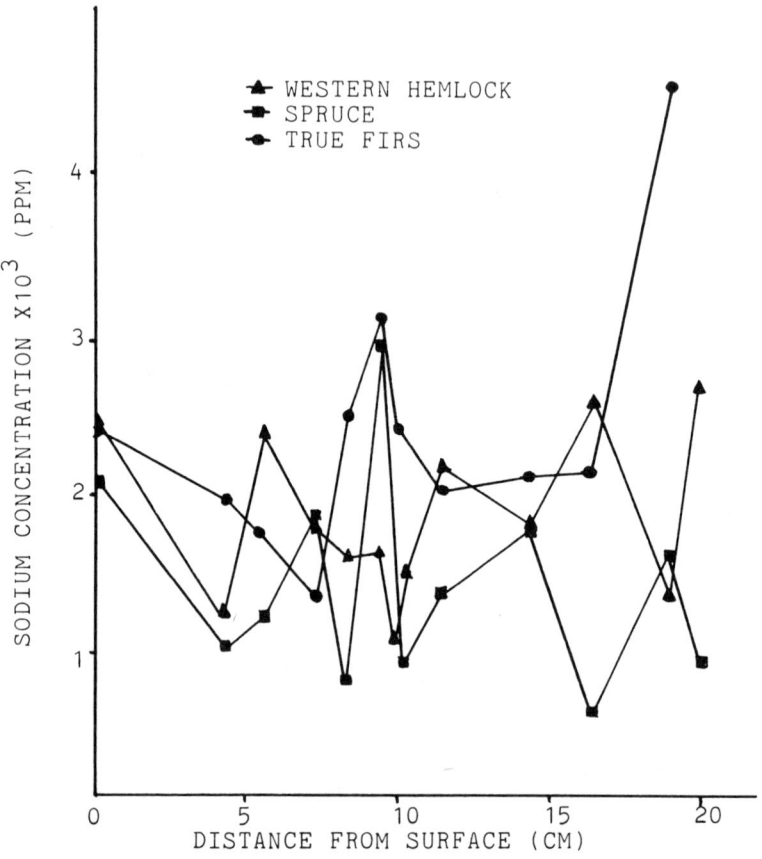

FIGURE 1. Profile of sodium concentration by distance from log surface.

values of 332 ppm for western hemlock. This was comparable to the potassium values for the beached log material. Edmonds did not determine sodium content. Potassium was not sifnificantly different for sitka spruce but was slightly higher for the fir.

To determine the effect of sodium concentration combined with log storage time for logs stored in saltwater, an additional sample of western hemlock logs with bark was obtained from Simpson Timber Company in Shelton, Washington. In addition to sodium values, potassium, calcium, and magnesium values were also determined as shown in Table II. It can be seen from these values that the majority of inorganics are

TABLE II. Inorganic Content of Western Hemlock Logs Stored in Puget Sound

Position in log	n	Na (ppm)		K (ppm)		Ca (ppm)		Mg (ppm)	
		$\bar{x}$	σ	$\bar{x}$	σ	$\bar{x}$	σ	$\bar{x}$	σ
Bark	20	10,714	4,264	22	44	4,376	2,642	1,456	497
Log surface[a]	30	3,466	3,972	23	47	551	336	392	318
10 cm depth	30	679	1,514	23	47	647	301	172	111
20 cm depth	30	308	290	30	54	822	670	178	98
30 cm depth	20	434	400	44	67	785	357	265	61

[a] Wood surface at the bark-wood interface.

contained within the bark, but they have all penetrated into the log. The logs were stored for periods ranging from 11 to 28 days. This information is used to supplement that for the southeast Alaska logs.

Figure 2 shows the profile of sodium concentration for logs stored at 11, 16, 20, and 28 days. A significant difference was found between bark content and surface content for all values combined but not for interior values. There was not a statistically significant difference between values at a given location. The logs stored for 28 days contained less sodium in the bark and at the wood surface than those stored for 20 and 16 days, respectively. Logs stored for 28 days contained higher levels of sodium at the 10-, 20-, and 30-cm locations than logs stored for lesser periods of time.

IV. HIGHER HEATING VALUES OF BEACHED LOGS

Higher heating values were determined for ovendried samples in a Parr Adiabatic Calorimeter using ASTM procedures. Results are given in Table III. Values determined for sitka spruce and Pacific silver fir were significantly higher than those in the literature. The value for western hemlock is in agreement with published values. There is no explanation for this. Variability in the value for sitka spruce may be due to differences in location. The difference for the Pacific silver fir values are probably due to species classification in the literature. Values used for white fir were those reported by Ince (1979). They are probably an average value for the true fir species mix. There was no significant difference for heat contents between surface and pit regions.

V. PROXIMATE ANALYSIS OF BEACHED LOGS

The proximate analysis was done using a standard coal furnace. The temperature was kept constant at 950°C ± 20°C using a 1-minute flaming time and a 6-minute soaking time. The sample consisted of one gram of material ground to a minus two-plus one millimeter particle size. Results are given in Table IV.

There is very little published information on proximate analyses for these species. Most values in the literature are for the bark portion only and, therefore, do not apply here. Mingle and Boubel (1968) determined values for western hemlock and white fir. Again, using the values published for white fir as a comparison for Pacific silver fir, beached western

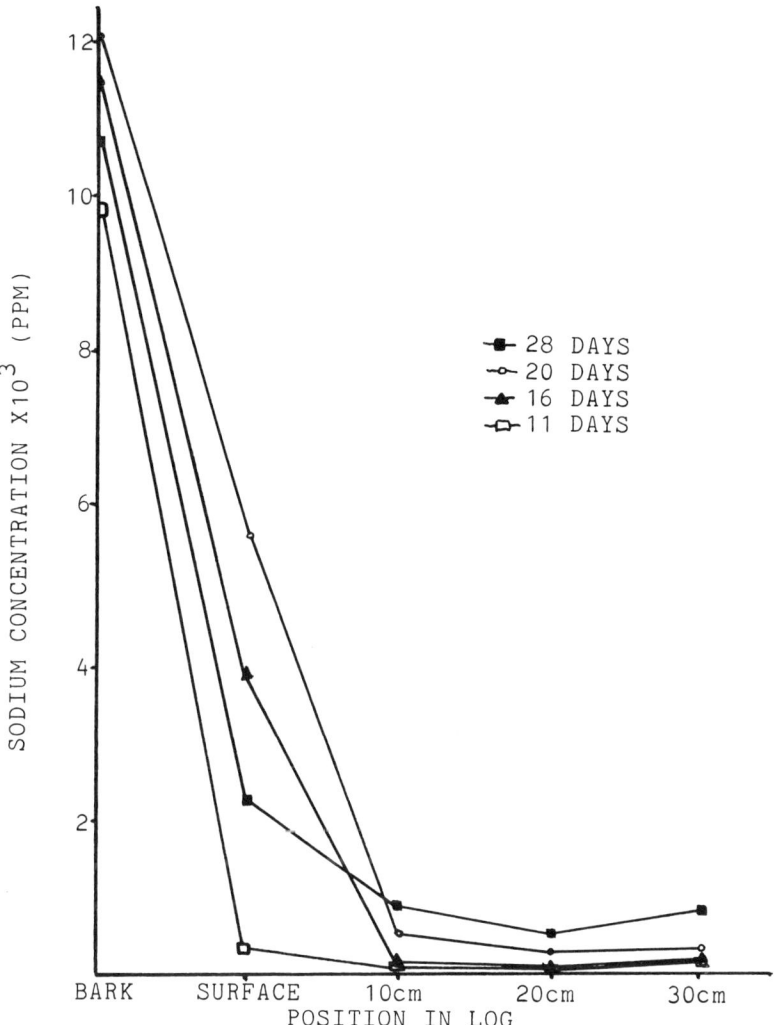

FIGURE 2. Profile of sodium concentration by log storage time.

TABLE III. Higher Heating Value

Species	Sample number	Average (Calories per ovendry gram)	Standard deviation (Calories per ovendry gram)
Western Hemlock			
Literature[a]	11	4863	248
Surface	56	4792	149
Pith	57	4852	141
Sitka Spruce			
Literature	1	4500	-
Surface	17	4753	72
Pith	17	4826	161
Pacific Silver Fir			
Literature[b]	2	4528	118
Surface	15	4748	79
Pith	15	4802	127

[a] Values obtained from unpublished data at the University of Washington and averaged with values given in Ince (1979).

[b] These values represent white fir values given in Ince (1979).

hemlock and Pacific silver fir have a significantly greater percentage of volatiles and less char and ash than those reported by Mingel and Bougel. The slight difference in ash content conforms with the increase in salt content; however, the reason for the difference between the others is not known. The same range in values was found for sitka spruce but comparable literature values could not be located. The values obtained also showed slightly greater char production for the pith material as compared to the shell material, but not significantly.

TABLE IV. Proximate Analysis

		% Volatiles		% Char		% Ash	
		(Percent of Ovendry Weight)					
Species	n	$\bar{X}$	σ	$\bar{X}$	σ	$\bar{X}$	σ
Western Hemlock							
Literature[a]	1	84.8	-	15.0	-	0.2	-
Surface	57	81.9	2.3	18.3	5.1	0.4	0.2
Pith	57	80.5	2.3	18.9	2.0	0.5	0.3
Spruce							
Surface	17	82.7	2.2	16.0	4.2	0.3	0.1
Pith	17	82.3	2.9	17.3	2.8	0.4	0.3
Pacific Silver Fir							
Literature[a]	1	84.4	-	15.1	-	0.5	-
Surface	15	81.9	3.1	17.5	2.8	0.5	0.3
Pith	15	80.1	2.4	19.1	2.1	0.7	0.5

[a] Mingle and Boubel, 1968.

VI. EFFECTS ON INDUSTRIAL OPERATIONS OF USING BEACHED LOGS

The effects of these findings on industrial operations depend on the specific type of operation. Obviously, corrosion on fuel and ash handling and transporting equipment will be present as well as corrosion and fouling of internal surfaces, as discussed previously. Equally important is the effect salt will have on the particulate collection equipment.

Given the previous results, western hemlock removed from the southeastern Alaskan beaches is expected to contain 9.75 kilograms of NaCl per metric tonne of material on the average, sitka spruce 8.44 kilograms, and true firs 10.69 kilograms. If it is assumed that 75% of the NaCl will exit through the boiler stack, then the amount to be collected per metric tonne input would be 7.31, 6.33, and 8.02 kilograms, respectively.

The total impact would depend on the type of operation as mentioned above. For example, consider a lumber drying operation that uses a wood-fired boiler to dry 70,800 cubic meters of Douglas-fir per year from an average of 75% moisture content (M.C.) (ovendry basis) to 19% M.C. (ovendry basis).

Using the values in Table V, which were calculated using a higher heating value of 4,796 calories per ovendry gram (an average for all three species), a net heating value of 1,457 calories per gram of fuel input can be realized. Over 17.8 million kilograms of water would have been evaporated from the lumber at 890 calories per gram of water evaporated. This would require 5,441 metric tonnes of fuel containing 52,396 kilograms of NaCl, if all species are averaged. From this 52.4×10^3 kg, 39,297 kilograms of NaCl would have to be collected as particulate matter.

A 50-MW power plant burning salt-loaded fuel would have a much greater NaCl production. A plant of this size would require approximately 42 metric tonnes of fuel per hour which would pass 7,289 kilograms of NaCl out of the stack per day, or 1,330 metric tonnes of NaCl per year. This would be in addition to 444 metric tonnes deposited in the ash stream.

Plants set up to handle salt-contaminated fuel would probably not have any difficulties utilizing this material as a fuel source. Those that begin to burn this material will have problems until noncorrosive materials are used in exterior handling systems and sufficient air particulate collection devices are installed. Because of the slagging problem, furnace temperatures will also need to be monitored closely.

TABLE V. Calculated Boiler Losses per Gram Fuel Input

Heat loss	Quantity (Calories per gram)
Stack heat loss due to moisture	332
Stack heat loss due to H_2 in fuel	179
Stack gas heat loss	333
Heat loss through furnace (4%)	96
Total heat loss	941
Recoverable heat	1457

VII. CONCLUSIONS

Wood fuels have the advantage over other solid and fossil fuels in that they do not create as many air pollution problems. Both coal and oil contain sulfur and both burn at higher temperatures resulting in emissions of sulfur and nitrogen oxides. Fly ash has posed problems for all fuels, but has been minimized by using efficient combustion techniques and stack gas cleaning equipment, such as mechanical collectors and electrostatic precipitators.

Wood fuels are also not as naturally corrosive to furnace and boiler components as the fossil fuels. Fossil fuels containing 25 ppm vanadium, 2.5% sulfur and 70 ppm sodium have been shown to combine in various corrosive forms (Babcock and Wilcox, 1978) and, depending on furnace conditions, either high temperature or low temperature corrosion can result. Wood fuels do not normally contain these elements.

Beached logs, however, do contain high levels of sodium which can have detrimental effects. Because the sodium chloride, which can result in flyash is submicron in size, it is very difficult to capture in collectors. An increase in stack emission and stack capacity can result.

Corrosive conditions could be set up inside the boiler with the formation of Na_2SO_4 and HCL which would degrade breechings as well as other components. Greater deposits would also be formed by reduced ash fusion points, decreasing overall boiler efficiency. Problems may be minimized with various additives and mixing with noncontaminated fuels to reduce the corrosive levels; however, it is not known what levels would have to be obtained.

REFERENCES

Babcock and Wilcox Company. 1978. "Steam, Its Generation and Use." The Babcock and Wilcox Company, New York.

Brady, J. D., and Jenkins, H. N. 1980. Wood Energy Emissions Control Technologies. *In* "Energy Generation and Cogeneration from Wood." Forest Products Research Society, Madison, Wisconsin.

Goldberg, H. J., and Bennett, R. P. 1980. Controlling Fireside Sodium Corrosion in Large Oil-fired Steam Generators. *Power 124*(1):73-75.

Ince, P. J. 1979. "How to Estimate Recoverable Heat Energy in Wood or Bark Fuels." U.S.D.A. For. Ser. Gen. Tech. Report FPL 29.

Leman, M. J. 1975. Special Environmental Problems Originated by Burning Bark from Saltwater Borne Logs. *In* "Wood and Bark Residues for Energy" (compiled by S. E. Corder). Proceedings of a conference held May 31, 1974. School of Forestry, Oregon State University, Corvallis.

Mingle, J. C., and Boubel, R. W. 1968. Proximate Fuel Analysis of Some Western Wood and Bark. *Wood Science 1* (1):29.

AN ASSESSMENT OF THE COSTS AND BENEFITS OF RECOVERING LOGGING RESIDUE FOR ENERGY USE

Ellen J. Hall

Envirosphere Company
Bellevue, Washington

I.	INTRODUCTION AND SUMMARY	219
II.	BACKGROUND	220
III.	AVAILABILITY OF LOGGING RESIDUE AND THE VALUE OF ENERGY	222
	A. Physical Availability of Logging Residue . . .	222
	B. Feasibility of Using Wood Fuel in Large Electric Generating or Cogenerating Facilities .	223
	C. Feasibility of Small Heating and Electric Generating Facilities	224
	D. The Value of Full Avoided Cost	225
IV.	THE COST OF RECOVERING LOGGING RESIDUE	227
	A. The Green Mountain Study	227
	B. Who Pays the Cost of Recovery?	228
	C. Reducing Recovery Costs Paid by the Purchaser .	230

V.	DETERMINING THE VALUE OF UNPRICED BENEFITS . . .	231
	A. Net Present Value vs. Net Public Benefits . .	231
	B. Evaluation of Hydroelectric Development in Hell's Canyon	232
	C. Evaluation of Timber Management Policies . .	233
VI.	BENEFITS OF RESIDUE REMOVAL	234
	A. Energy Production	234
	B. Smoke Management	237
	C. Soil Disturbance	241
	D. Improved Ground Access and Earlier Reforestation	242
	E. Reduced Risk of Wildfire	243
	F. Reduced Burning Cost	244
	G. Reduced Planting Cost	244
	H. Results of the Analysis	245
VII.	REDUCING THE COST OF YUM YARDING	246
VIII.	SITE SPECIFIC DECISION MAKING	247
IX.	SUMMARY	250
X.	NOTES	251
XI.	APPENDIX	252

I. INTRODUCTION AND SUMMARY

The intent of this chapter is to outline a procedure for evaluating the benefits and costs of recovering logging residue for energy use. The cost of recovering the wood remaining after a timber harvest is frequently greater than the value of that wood as an energy source. Thus, much of the material is left in the woods to rot, or is burned when the site is prepared for planting a new stand of trees. Suggestions have been made that when selling timber from the National Forests, the U.S. Forest Service should reduce the price a purchaser must pay for the merchantable wood and, in effect, subsidize the removal of the "unmerchantable" wood. Given such a situation, the material could be made available as fuel at a price buyers would be willing to pay.

The question remains as to whether it is in the best interest of society for the Forest Service to pursue such a course. While such a policy would reduce the revenue received from timber sales, and thus reduce the amount of money the Forest Service returns to the U.S. Treasury, it would also provide other benefits which are not normally considered in the timber sale decision. Those benefits include such things as reduced air pollution from slash burning, reduced costs for establishing and managing future stands, and reduced wildfire risk. This chapter presents a method which can be used to determine if, in fact, it is in the public interest for the Forest Service to pay for the removal of logging slash so that it is available for use as fuel. It should be noted that such material might also be put to other uses. For the purposes of this chapter, the assumption is made that recovered logging slash reaches the fuel market.

The major finds are as follows:

(1) The demand for wood as an energy source is derived from the demand for final products: electricity, process steam for drying plywood, and so on. The current price of electricity in the Pacific Northwest, defined to include Oregon, Washington, and northern Idaho, limits the price that can be paid for fuelwood to generate electricity.

(2) The price of alternative fuels also places a limit on the price that potential customers can pay for logging residue. Logging residue cannot be sold for $20 per ton if mill waste is readily available at $15 per ton.

(3) The cost of yarding unmerchantable material (YUM) is the major factor in determining the overall cost of providing logging residue for fuel. Cost estimates vary widely, depending on the site.

(4) The price at which logging residue can be profitably sold is lower if the Forest Service pays for its removal by accepting a lower price for merchantable timber on the sale.

(5) There is a method to determine if it is worthwhile for the Forest Service to pay for the removal of wood for energy. The procedure follows a 4-step process:

(a) Determine all relevant costs and benefits. Prices and costs are estimated for as many factors as possible.
(b) Calculate the net present value of each management alternative, that is, determine the value of benefits minus cost.
(c) If the costs exceed the benefits, compare the net cost to the remaining unpriced benefits.
(d) Make a subjective judgment about whether the remaining benefits are worth the cost of acquiring them. Where it is feasible to do so, compare the cost of acquiring these benefits with the cost of acquiring similar benefits through other means.

(6) More research is needed to provide better estimates of the dollar value of benefits. Specific areas needing more research include the effects of intensive yarding on soils, the effect on wildfire risk, and the effect on subsequent management costs.

(7) Following this analytical procedure on the Green Mountain sale indicated that the cost of smoke reduction was somewhat higher than that found in other industries. The evaluation of other sites could produce different results.

(8) The cost of emission control on the Green Mountain sale could be reduced if yarding costs were reduced by the use of a miniyarder for the smaller material.

(9) Results would be even better if the removal of logging residue could completely eliminate the need to burn harvested areas in order to dispose of the remaining slash and prepare the site for planting.

(10) The real answer to the question of whether or not residue removal is worthwhile can only be determined on a site specific, project specific, basis. The Forest Service should have the flexibility to make those determinations.

II. BACKGROUND

The energy crisis of the 1970s fueled a renewed interest in the use of wood for fuel. This in turn led to an increasing interest in biomass recovery. In 1978, oil was $33 per barrel and coal was $38 per ton (BPA, 1983). Prices showed every sign of continuing to rise. Utilities in the Northwest were predicting substantial growth in electric loads for the foreseeable future. The time appeared ripe to find a way to use formerly "useless" materials to plug the widening energy gap. Research was undertaken to determine the quantity of wood available for energy—wood from logging residue, road slash, mill residue, harvest of undesirable species such as alder, and biomass plantations begun for the express purpose of providing wood for fuel.

Between 1978 and 1982, however, the situation changed dramatically. The growth rate in the price of nonrenewable fuels abated, and some prices started to drop. The price of coal, for example, dropped to $28 per ton. New projections of demand for electric energy were far lower than earlier forecasts. Simultaneously, early research efforts were finding that there are numerous economic and institutional barriers to the use of wood as a fuel. Research into these issues continues, both because of the long term nature of many of the projects and because many people have come to realize that wood energy has good long term potential.

Throughout this period, additional research needs were being identified. A report by the Wood & Biomass Subcommittee of the Pacific Northwest Utilities Conference Committee (PNUCC, 1982) included the following among its recommendations:

> The Subcommittee, recognizing the need for innovation in procurement of wood fuels, endorses the principles of intensive forest management practices throughout the Region. The Subcommittee recommends that the U.S. Forest Service, as a lead agency in this area, consider changing its procedures as an incentive to encourage residual utilization in order to accommodate:
> (1) Crediting forest management benefits in determining timber sale prices;
> (2) Encouraging full development of the existing timber resource to multiproduct markets;
> (3) Using tree-measurement sale methods, rather than log scaling, in order to stimulate forest residue removal; and
> (4) Preserving piled logging residues for nearterm use.

The forest management benefits referred to in Item (1) include, but are not limited to, reduced slash disposal cost and reduced reforestation cost. The implication is that the inclusion of such benefits in timber sale decisions would work in favor of greater wood utilization.

The purpose of this chapter is to address this point not in an institutional sense, but in an economic sense. Thus, the question asked is not "How feasible or desirable is it to change the timber appraisal procedure?" The questions asked are "Could the consideration of nontimber benefits in timber sale decisions foster greater utilization of wood for energy?" Would such a policy increase the net benefits to society?

These questions can be considered within the context of the agency's general approach to timber sale decisions. The usual approach explicitly considers residue conditions and disposal options in the harvesting decision. Timber cutting, logging, transportation, slash treatment and regeneration are all part of the same decision, not separate job-by-job decisions. While the analysis presented here focuses on only a part of the costs and benefits included in a timber sale, the same analysis could be incorporated in an overall system evaluation of a timber sale.

III. AVAILABILITY OF LOGGING RESIDUE AND THE VALUE OF ENERGY

Most of the research on the topic of residue recovery has concentrated on the physical availability of logging residue, the cost of making material suitable for use as fuel available to the market, the feasibility of using wood fuel to support 15 to 50 megawatt (MW) electric generating facilities, and the feasibility of using wood fuel to support smaller (e.g., 2-15 MW) cogeneration plants or boilers. A brief review of their findings will help define the nature of the problem.

A. Physical Availability of Logging Residue

Some of the early research on the topic of residue recovery produced estimates of the physical quantity of logging residue available. Grantham and Howard (1980), for example, estimate the quantity of logging residue available in the Pacific Coast region at 1,038,100 thousand cubic feet, or 14,430,000 oven dry tons. This estimate includes all wood and bark above a one foot stump and 4 inches or larger in diameter. The amount of new material which becomes available each year is a function of the number of acres harvested and the age and species composition of the harvest. Logging residue is expected to decrease over time, primarily because the average age of harvested trees is declining. Younger, second growth trees produce much less residue per acre than old growth stands. Howard (1979) notes, however, that the main concerns are not the total volume of material in the forest, but the cost and future availability of the material.

Both cost and future availability are to some degree a function of local, site specific circumstances rather than broad regional totals. Recent research has tended to focus more on information useful on a site specific basis. Howard, for example, presents information on volume by substate area, land ownership and cutting type (1981a), and by characteristics such as diameter, length, and distance to a road (1981b). Keegan's (1981) assessment of the volume of residue available in eastern Montana and north Idaho includes a definition of supply area characteristics for three distinct processing centers.

Increasingly, studies which seek to define the quantity of logging residue available for energy uses mention the difference between physical availability and economic availability. This is because they recognize the fact that the demand for wood as fuel is derived from the demand for heat, steam, and electricity. To be competitive with other sources of energy such as coal and oil, wood must be convertible to energy at a competitive price.

B. *Feasibility of Using Wood Fuel in Large Electric Generating or Cogenerating Facilities*

When the energy crisis was at its worst, much interest was shown in analyzing the feasibility of using wood as the primary fuel for large electric generating facilities (Rocket Research, 1980; Nor'west-Pacific, 1981). The Pacific Northwest Utilities Conference Committee (1982) estimates that each year we create approximately 2.5 million oven dry tons (ODT) of logging slash and 0.3 million ODT of unused mill residue which is practically recoverable for fuel uses in the Northwest region. That equals an electrical generation potential of about 425 MW using central station generation technology. Their estimate of capital costs for central station wood-fired plants in 1982 dollars are $1491/kilowatt and $1278/kilowatt for 25 and 50-MW plants, respectively. This compares to a capital cost of $1033 per kilowatt for new coal fired thermal plants (NPPC, 1983). The price of fuel becomes a critical factor.

The Northwest Conservation and Electric Power Plan (NPPC, 1983) estimates the current cost of producing electricity in the northwest is 2.5 cents per kilowatt hour. The estimated cost of energy from a new thermal coal plant would be about 6.7 cents/kwh while conservation could be acquired at 1.8 cents per kilowatt hour. Estimated cost for a wood fired plant near Eugene, Oregon, was 6.4 to 11.3 cents per kilowatt hour, with fuel estimated at $16.50 per ton chipped and delivered to the plant (Nor'west Pacific Corp., 1981). Even the lower price of 6.4 cents per kilowatt hour is considerably more than the present price of electricity in the area, and only marginally better than the price of energy from coal.

Another study (Rocket Research, 1980) evaluated the feasibility of establishing a wood- and refuse-burning 25 MW thermal/electrical cogeneration facility in eastern Lewis County, Washington. They estimated that a public utility ownership could produce energy for approximately 4.0 cents/kwh, with fuel from logging residue estimated at $10 to $31 per dry ton. The price per kilowatt hour is higher than current electricity prices, but within the range of other alternatives for future generation.

The high cost of producing electricity from a wood-fired system does not totally preclude its use. Washington Water Power has recently submitted a request to have the price of generating electricity from its new Kettle Falls plant incorporated in its rate base. The fully allocated cost of Kettle Falls is 7.9 cents per kilowatt hour. If approved by the public utility commission, that cost will help determine the average price paid by Washington Water Power's customers.

Beside the price of fuel, another major problem faced by large facilities is the uncertainty of log term fuel supplies. Bergvall and others (1978) note that the demand for all wood products has been increasing steadily. Material that was once left in the woods is now being profitably used to manufacture paper, particleboard, and fabricated lumber. Continuing technological advances will further reduce the quantity of residual material available for energy. A further problem is the difficulty utilities might find in getting long term contracts with fuel suppliers. Because logging residue is just that, the leftover material from a timber harvest, the boom and bust nature of the timber market has a similar cyclical effect on the volume of new residue being produced. Solutions to these problems include the entrance of third person fuel supply contractors in the market, and the continued piling of residue from timber sales regardless of the short term outlook for its use (PNUCC, 1982).

C. *Feasibility of Small Heating and Electric Generating Facilities*

Another approach to using wood as fuel is to direct-fire small boilers for heating. These can be small enough to heat a single building or complex of buildings. The State of Washington, for example, is studying the feasibility of converting some of its coal burning boilers to accept wood as fuel (Juhasz, 1983). This can be particularly cost effective if the existing boiler cannot continue to burn coal and still meet EPA air quality standards. Converting to wood is cheaper than replacing or upgrading the equipment to meet today's standards. The long term success of such a venture is dependent on the relative cost of fuel. Johnson (1982) has reported on an operationally successful test of wood-coal combination firing in a boiler originally designed for coal. At $38 per ton for coal and mill residue available for little or no cost, the project proved worthwhile. When coal dropped to $28 per ton and wood fuel rose to $20 per ton, wood fuel lost its economic advantage.

The small boiler can supply process steam as well as heating steam and can be used to simultaneously supply steam for electricity generation. A recent boon to private developers of generation and cogeneration facilities is the Supreme Court decision on controversial federal rules governing electricity sales between small power producers and electric utilities (Morgan, 1983). The Supreme Court decision overturned a lower Appeals Court decision which restricted the ability of independent power producers to sell power on utility grids. Utilities must now agree to buy power from small producers at rates based on "full avoided cost," or the cost of electricity from new conventional sources. They must also guarantee independent generators the right to interconnect with the utilities' power grids. The purpose of the Public Utility Regulatory Policies Act of 1978 (PURPA) is to encourage the use of cogeneration and small power production.

D. *The Value of Energy at Full Avoided Cost*

Full avoided costs currently published by northwest utilities are running from 1.0 to 11.0 cents per kilowatt hour, depending on the length of the contract, the type of energy, and other contract specifications. Small power producers, including those using wood fuels, have an opportunity to negotiate with utility companies and with Bonneville Power Administration to sell any electricity which they produce. Utilities are generally willing to pay a higher price for a long term contract, locking in a price which is higher than average now but could be cheaper than other alternatives over the life of the contract. Idaho Power Company, for example, recently signed a 35 year contract to purchase the output of Evergreen Forest Products' proposed five megawatt cogeneration facility. The Idaho Public Utility Commission set the power purchase price at Idaho Power's full avoided cost of 7.2 cents per kilowatt hour (Bioenergy Bulletin, 1982). Table I displays the range of relevant electric generating costs.

TABLE I. Electric Generating Costs in the Pacific Northwest[a]

Generating method	Cents per Kilowatt hour
Low range - full avoided cost	1.0
Conservation	1.8
Current system mix	2.5
Coal fired thermal	6.7
High range - full avoided cost	11.0

[a] Sources: Conservation, current mixed system, and coal fired thermal (Northwest Power Planning Council, 1983). Full avoided costs (Pacific Power and Light, 1982; Washington Water Power, 1983; Puget Sound Power and Light, 1983).

Another option which exists is the sale of Pacific Northwest energy to utilities in other parts of the country. Current negotiations between Bonneville Power Administration and California utilities hold some promise (Dorn, 1983). The cost of California's gas- and oil-fired thermal energy is about 10 cents per kilowatt hour. If BPA or northwest utilities can generate electricity for less and sell to California, both buyer and seller will benefit. This is another demonstration of the importance of the energy market in determining the use of wood as fuel. As Bergvall (1978) noted, the viability of power production from wood is up to the consumer. As the price of energy rises, unconventional fuel sources become exploitable.

A common thread running through all the above mentioned studies is the cost of acquiring fuel. Logging residue, or logging slash, is what remains on the ground after a timber purchaser has removed all the desirable "merchantable" material. The cost of acquiring the logging slash for fuel might include a payment for the volume of biomass recovered and the cost of recovery—getting the material to the road, loading it on a truck, delivering it to its destination, and chipping it into pieces suitable for fuel. The elements affecting recovery cost include terrain, the equipment used, the stand age, size of the material being moved, crew skills and morale, and general operator efficiency. Costs are also affected by specifications in the Forest Service contract.

IV. THE COST OF RECOVERING LOGGING RESIDUE

The cost of providing logging residue for energy use varies greatly. Studies made in the last five years have produced estimates (all converted to 1982 constant dollars) of $27 to $99 per oven dry ton, chipped and delivered (Kierulff and Adams, 1980; Brown, 1982). This is roughly equivalent to $13.50 to $45.50 per green ton. To put this cost in perspective, mill residue is generally available for less than $30 per dry ton.

Given the variability in the cost of recovering logging residue, this chapter cannot, and does not, claim to represent the definitive cost of recovery, or even the "average" or "representative" recovery cost. Instead, it presents only the process by which costs are determined, the process by which benefits can be estimated, and the procedures which can be used to compare the two.

A. *The Green Mountain Study*

An extensive study of residue removal has been made for the Green Mountain timber sale on the Willamette National Forest (Brown, 1981). The Green Mountain sale was part of a wood utilization study sponsored by Oregon Governor Victor Atiyeh's Wood Residue Utilization Committee and the USFS Willamette National Forest. The USFS Pacific Northwest Forest and Range Experiment Station assisted with the project design. The study purpose was to examine the economic feasibility of removing (yarding) logging slash and hauling that material to a location where it could be processed and/or made available for the generation of energy. They sought to demonstrate that a market does or may exist for the use of wood residues for energy production.

The contract on the Green Mountain sale required the purchaser to YUM (yard unmerchantable material) or PUM (pile unmerchantable material) to certain specifications. YUM is not always required on Forest Service timber sales, but is one of several methods used to dispose of the slash left after harvest and to prepare the site for planting a new stand of trees. Another common method is broadcast burning, which entails burning the entire harvested area. Both methods are sometimes used in combination.

Normal contract specifications for YUM require a timber purchaser to remove all logging slash that is at least 8 in. in diameter and 10 ft long. Two units of the Green Mountain sale were logged to this specification and acted as control units. On the remaining four units, logs and pieces down to

6 in. diameter and 6 ft long were removed, and wood chunks larger than 16 in. × 2 ft were also removed. For ease of presentation, the latter units will be referred to throughout this chapter as being yarded to 6 in. × 6 ft specifications.

Careful records were kept on yarding, loading, hauling and other aspects of the sale. Units yarded to the 8 in. × 10 ft specifications cost the logging contractor an average of $21.37 per green ton to yard and load the unmerchantable material and haul it 60 miles. At an average recovery of 90 tons of unmerchantable material per acre, the cost per acre was $1923. The equivalent figures on the units yarded to smaller specifications are $26.14 per green ton or $2981 per acre, given 114 tons per acre average recovery. Compared to a no-YUM situation, working to 8 in. × 10 ft specifications costs an additional $1923 per acre. Going to the smaller specification adds $1058 per acre to that cost.

B. Who Pays the Cost of Recovery

To understand who pays this additional cost, one must understand the way that the Forest Service sells timber. Brown (1982) has summarized the procedure as follows. The Forest Service auctions its timber as stumpage. The agency uses a standard appraisal system to estimate the minimum price it will accept for the stumpage. Minimum stumpage value is equal to the estimated final value of the end products less the estimated logging and processing costs. As part of the estimated logging cost, the Forest Service includes the cost of yarding or piling unmerchantable material (YUM or PUM). The Forest Service generally handles the remaining slash disposal work, primarily broadcast or pile burning. The purchaser makes a brush disposal (BD) deposit to cover the Forest Service estimate of the cost of the disposal work plus overhead.

In the absence of competition for Forest Service stumpage, we would expect the purchaser to pay the minimum appraised value. If a contract required yarding to smaller specifications, the logging costs would go up and the stumpage value would go down by an equal amount. Assuming the logging contractor could do the work for exactly the cost estimated, he might be indifferent to the amount of work required (Note 1). His out-of-pocket costs would be the same either way.

The situation is slightly less straightforward when the timber market is more competitive. Timber sales are generally bid up far higher than the minimum appraised value. On the Green Mountain sale, for example, the timber appraised at $124.20 per thousand board feet (MBF), including stumpage, road maintenance, and slash disposal deposits. The actual bid on the Green Mountain sale was $358/MBF. Purchasers are able to bid more than the minimum for timber for several reasons. They may believe that the final product value is higher than the Forest Service estimate. They may be able to do the logging and/or processing for less than the appraisal estimate. They may also be willing to take less profit, or even a loss, in order to secure a certain volume of timber supply at a given time.

Under competitive conditions, the one-to-one relationship between logging costs and stumpage value gets lost. One can still assume, however, that the purchaser is taking all costs into account when making a bid. For that reason we can assume that when smaller yarding specifications are written into a timber contract, the Forest Service pays the additional yarding cost in the form of lower stumpage value. Brown (1982) notes that the Fuel Treatment Appraisal allows for slash disposal by "Removal of Piles or Decks with or without the Sale of Products." This means that loading and at least partial hauling costs of unmerchantable material can also be covered by the appraisal.

Theoretically then, it is the Forest Service that pays the cost of any additional yarding required by the timber sale contract. The Forest Service pays in the form of a lower price received for stumpage. In the long run, it is the nation's taxpayers who pay the cost because they receive a lower cash return on their investment in national forest land management. The compensating benefits for this loss will be the topic of the next section.

It is important to note at this point that the Forest Service is not at liberty to require YUM yarding for the sole purpose of making the material available for energy or other uses. In FSM 2403.25, the Forest Service Manual (USDA, 1977) states that "As a general rule, the total appraised value of a tract of timber will not be reduced to obtain utilization of a species, size, or class material. . . . An exception to the rule prohibiting a reduction in tract value may be made when the leaving of a species, size, or class material will result in an increase in the cost of slash disposal necessary to achieve acceptable standards. In such cases, utilization may be required and the tract value may be reduced by an amount not to exceed the increase in disposal cost which otherwise would result." For example, the tract value could be reduced by up to $200 per acre if it would save an additional $200 required by the next best alternative disposal method.

In addition, the manual states that "utilization standards, requiring the removal of whole pieces usable only for fiber products, must be established to reasonably balance required removal with demand. . . . As a general rule, established standards should not be changed so as to require removal of smaller or more defective material until it is demonstrated on the basis of actual experience that the additional material can be manufactured and sold successfully and that any new equipment necessary to do so is readily available to purchasers." This direction is provided to protect both the Forest Service and the purchaser from incurring unnecessary costs of recovering material which has no market. This direction might run counter to contract specifications being proposed here.

C. Reducing Recovery Costs Paid by the Purchaser

As the foregoing suggests, the purchaser or logging contractor need not sell recovered material at the full price required to cover his costs of yarding, loading, and hauling the material if the costs of those activities is covered in the timber sale appraisal (that is, the costs are effectively paid for by the Forest Service). Brown (1982) presents an illustration of that point, based on data from the Green Mountain sale. He presents three cases, all of which use cost data from unit 3 of the sale and allow the purchaser a 15% profit on his added costs as shown in Table II.

Case A: This case assumes normal contract specifications, i.e., all material larger than 8" × 10' must be yarded to the landing. Smaller material may be left on the forest floor. The costs for falling and bucking all material is part of the total falling and bucking cost for merchantable material. It is therefore part of the purchaser's fixed costs charged against the merchantable timber. Because the contract calls

TABLE II. Recovery Costs of Unmerchantable Material on Green Mountain Sale Unit 3[a]

Activity	>6"×6' & <8"×10'	>8"×10'
	(Dollars per green ton)	
Fall and buck	1.94	1.35
Yard (Skyline)	26.98	9.37
Load	9.47	2.79
Haul (60 miles)	6.44	6.14
15% profit	varies	varies

[a] Source: Brown (1981) updated to 1982 constant dollars.

for yarding all material greater than 8" × 10', that cost has also been reflected in the purchaser's bid for the merchantable timber. Delivering the larger residue to a buyer 60 miles away, the purchaser would have to charge $10.27 per green ton to recover his additional costs and make a 15% profit (($2.79 + $6.14) × 1.15). In this case, the cost to yard the material smaller than 8" × 10' and larger than 6"× 6' is not covered by the appraisal. For the smaller material, therefore, he would have to charge $49.32 per green ton (($26.98 + $9.47 + $6.44) × 1.15) to recover costs plus make a 15% profit.

Case B: In this case, the contract would specify that all material larger than 6" × 6' must be yarded to the landing. The purchaser would calculate the additional cost of yarding the small material and reduce his bid for the merchantable timber accordingly. All felling, bucking, and yarding costs are now part of the purchaser's fixed costs. He need recover only his variable costs of loading and hauling the material to a buyer. The price for the larger material would remain $10.27, as in Case A. The price for the smaller material could be reduced to $18.30 (($9.47 + $6.44) × 1.15) per green ton.

Case C: In the third case, the timber sale contract would require the purchaser to yard all material greater than 6" × 6' and haul it to a disposal site 10 miles away. Assuming the haul to the disposal site was $2.36 per ton, it would be that much less expensive to haul to a buyer 60 miles away. The market price of the wood fiber could be even lower. The larger material could be sold for $4.35 per ton (($6.14 - $2.36) × 1.15). The smaller material could be sold for $4.69 per ton (($6.44 - $2.36 × 1.15), and still return a 15% profit.

From the data presented by Brown, we can see that contract specifications calling for YUM and hauling to a disposal site could provide the timber sale purchaser with a reasonable profit and the energy wood buyer with wood fiber at an acceptable price. The question remains as to whether it would be worthwhile for the Forest Service to pursue such a course, since it would result in lower stumpage prices received for the merchantable timber.

V. DETERMINING THE VALUE OF UNPRICED BENEFITS

A. *Net Present Value vs. Net Public Benefits*

The National Forest Management Act of 1976 and implementing regulations (Federal Register, 1979) set the standard for evaluating forest land management plans. Each alternative plan must be evaluated for its net present value and net public benefits. As defined by the regulations, net public benefits is a measure of all costs and benefits of a plan, regardless of whether all components can be quantified or expressed in monetary terms. Net present value is a similar concept, but it is limited to those costs and benefits which have, or can be assigned, market values. The selected management plan is to be that which maximizes net public benefits. That choice is the plan which has the greatest excess of benefits over costs.

The somewhat nebulous nature of the net public benefits concept is acknowledged. One person's wilderness "benefit", for example, is another person's wilderness "cost." One logical test for the reasonableness of whether or not a given plan really maximizes net public benefits is to compare it to the plan which maximizes net present value. When an alternative includes costs to produce unpriced benefits such as cleaner air or more plant diversity, net present value is reduced. The loss in net present value due to the addition of unpriced benefits is termed the opportunity cost of providing those benefits. Evaluation can then take the form of this question: Does it seem reasonable that this set of unpriced benefits can be worth the opportunity cost of attaining them?

B. Evaluation of Hydroelectric Development of Hell's Canyon

While the National Forest Management Act applies only to forest land management plans rather than all the agency's projects and policies, the concept is appropriate to evaluate any public project. Krutilla and Fisher (1975) for example, used a similar technique to evaluate a proposed hydroelectric dam in Hell's Canyon. The first alternative considered was the construction of a hydroelectric dam which would inundate a large section of the scenic Hell's Canyon. Given the projected price of electricity and expected rate of technological advance, the authors estimated a net present value for this alternative of $9.8 to $18.5 million, depending on the discount rate assumed. The second alternative was to not build the dam—the do nothing alternative. That alternative would preserve the free flowing nature of the river and the grandeur of the canyon itself. Is it likely that these unpriced benefits are worth the cost of preserving them?

Krutilla and Fisher believed that the answer is yes. One of their reasons for reaching that conclusion is that natural environments are becoming increasingly scarce, therefore their value relative to other commodities is increasing. The authors' estimate was that the value of the "do nothing" alternative need equal only $40,000 to $150,000 in the first year in order to equal the present value of development. The range of estimates is determined by varied assumptions about the discount rate and the real rate of increase in the value of natural environments. Krutilla and Fisher conclude that recreation visits to the site could easily be worth $40,000 to $150,000 a year. Net public benefits would therefore be maximized under the nondevelopment alternative.

C. Evaluation of Timber Management Policies

Hyde (1980) used a similar method to examine the effect of three public timber management policies—even flow, sustained yield, and harvest at culmination of mean annual increment—on the value of standing timber and the value of long-term timber management (Note 2). Hyde used a case study approach on the French Creek drainage on the Willamette National Forest. The area has about 700 million board feet of standing timber, most of which is mature timber 100-400 years old. The area also offers unique recreational opportunities and has been considered for backcountry recreation or wilderness designation. Hyde first determined that the unconstrained net value of existing timber in the French Pete drainage is about $116 to $147 million, assuming a 10% discount rate and immediate harvest over a period of five years. This is equivalent to an annual value of $11.6 to $14.7 million forever.

Hyde next looked at the impact of certain policies on the net present value of the French Pete timber. When applied to French Pete, the policy of even flow sustained yield extends the liquidation period for existing stands to 241 years. Holding the timber for that time reduces the net value to less than zero: -$9.4 million. The land might be more efficiently allocated to nontimber uses such as recreation if it could be provided for a net cost of less than $9.4 million. Hyde's final conclusion is that social welfare might be better served by withdrawing the even flow policy and managing the timber according to efficiency criteria. It seems unreasonable to assume that the even flow policy could be worth its opportunity cost.

The empirical results of the preceding examples support several conclusions which are relevant to this research.

(1) It is possible to compare two management alternatives, one of which produces benefits with market values and one of which produces mainly unpriced benefits. The opportunity cost of providing the unpriced benefits provides a benchmark for evaluating their worth.

(2) Public policies can be similarly evaluated.

The groundwork is now laid for this analysis of salvaging unmerchantable material for energy use. Section III showed that the price of electricity and the price of alternative fuels puts a ceiling on the price that can be paid for wood fuel. Section IV covered the factors which determine the delivered price of wood fiber recovered from Forest Service timber sales. Next, Brown's argument that the Forest Service should, in effect, pay for the purchaser's removal of logging residue was reviewed. Section V established a procedure for deciding whether such a policy would be in the public interest.

The next step will be to examine the possible benefits which would be enjoyed by pursuing such a policy. The strategy will be to estimate values for some benefits and compare the remaining unpriced benefits to the cost of achieving them. In order to illustrate how the pieces fit together, data will be introduced from the Green Mountain sale and from other sources. The actual numbers used are applicable only to this example, but they help identify data sources and point out areas where more research is needed. Costs and benefits will be summarized on Table III. Derivation of costs per acre is shown on Appendix Table A-1. Following the analysis in Section VI, several remaining issues will be addressed.

VI. BENEFITS OF RESIDUE REMOVAL

Several benefits of residue removal are mentioned consistently in discussions of the subject (see, for example, Kierulff and Adams, 1980. The first benefit is, of course, increased energy production through the more complete utilization of resources. A second major benefit is a reduction in the smoke produced by slash burning. Other benefits include a reduced incidence of escaped fires, lower wildfire hazard, easier access to the land for later management, earlier reforestation, less soil damage, reduced burning costs, reduced planting costs, and general aesthetics. Each of these elements will be discussed in this section. Estimates of the value of each are summarized in Table III.

A. Energy Production

The value of wood as fuel is derived from the value of the end product it is used to produce. For example, the value of wood used in a cogeneration process in a sawmill is derived from the value of the lumber dried by the steam produced and

TABLE III. *The Implied Value of Smoke Standard and Intensive Yumming Compared to No YUM*

Parameter	Standard YUM >8"×10' (Dollars per acre)	Intensive YUM >6"×6'
Costs absorbed by Forest Service through lower stumpage value		
Yard	955	1748
Load	328	466
Haul to disposal site	213	270
Total Cost	1496	2484
Benefits		
Net energy value (equals delivered value/ton at mill less haul cost per ton from disposal site to mill times tons per acre)	924	1215
Smoke management cost avoided	0	0 to 14.02
Present value of earlier reforestation	0	0 to 17.00
Reduced risk of wildfire	0 to 13.10	0 to 26.20
Reduced burning cost	50	50
Reduced planting cost	50	50
Total priced benefits	1024 to 1037	1315 to 1372
Net Present Value	−472 to −459	−1169 to −1112
Physical smoke reduction	364 lbs	1024 lbs
Implied value per pound of smoke reduction	1.26 to 1.30	1.09 to 1.14

the value of the electricity generated. Profits are maximized when the difference between the cost of all input and the value of all output is maximized. Thus, the economic efficiency of the operation is enhanced when costs can be reduced relative to the value of the output. If all other costs are equal, the additional, or marginal, value of wood fuel is the difference between the cost of wood fuel and the cost of the next best alternative fuel. Where all other costs are not equal, such as when boilers must be converted from coal to wood burning, those costs must enter the equation.

To simplify the process of estimating the value of wood as fuel, we can use the price of mill residue as an estimate of the cost of the next best alternative fuel. Mill residue is already used extensively as a fuel in the forest industry, and the market is well established.

The price of mill waste tends to cycle with the lumber market. When the market is up and sawmills are active, the supply of mill waste rises relative to the demand for hog fuel or pulp chips. The price of mill residue is accordingly very low. When the lumber market is depressed, sawmill activity is reduced and the production of mill waste is correspondingly smaller. Competition for the decreased supply of material is keener, and the price rises. Frequently, paper companies desiring the residue for pulp chips can outbid those who would use the material for energy production. In any case, the price is higher for any purchaser.

Given the variable nature of mill residue prices, we will use an estimate that represents an "average" price. Brown (1981) cites prices of $16 to $28 per green ton ($18.88 to $33.05 in 1982 constant dollars). In 1982, Potlatch Corporation in Idaho was paying up to about $19 per green ton and Boise Cascade was paying up to $21. By way of contrast, Washington Water Power estimates that mill residue used to fuel its Kettle Falls power plant has a weighted average price under $12 per green ton (Anderson, 1983). Mill waste can be bought on the spot market at prices as low as $5 per ton. This analysis assumes a price of $15 per green ton.

The dollar value of using logging slash for fuel is equal to the difference between its delivered cost and the $15 cost of using mill residue instead (Note 3). For the cost of delivering logging slash, this analysis assumes a case similar to Case C on page 231. In that case, the Forest Service provides the maximum advantage to the purchaser by including all falling, bucking, yarding, and loading costs in the timber sale appraisal. Hauling costs to the disposal site are also included. This analysis will be the same as Case C except that it is based on all the units of the Green Mountain sale, not just on unit 3. The logging contractor (or agent) must then charge $4.73 per green ton to make a 15% profit on material delivered from units with standard 8" × 10' specification and $4.34 per green ton for material from the 6" × 6' specification units. Appendix Table A-II includes the calculation for these estimates. Subtracting each price from the alternative price of mill residue, the following net energy value of logging slash is estimated:

From 8" × 10' units: $15.00 - $4.73 = $10.27/green ton

From 6" × 6' units: $15.00 - $4.34 = $10.66/green ton

The amount of unmerchantable material removed from the two units did not vary greatly on a per acre basis. Units yarded to 8" × 10' produced an average of 90 green tons per acre. Units yarded to 6" × 6' produced 114 green tons per acre. Converting the energy values above from a dollar per ton basis to dollars per acre gives us:

From 8" × 10' units: $10.27/ton × 90 tons/acre = $924/acre

From 6" × 6' units: $10.66/ton × 114 tons/acre = $1215/acre

These estimates of net energy value per acre are shown on Table III. The estimates are sensitive to a number of assumptions. The first is the price assumed for the alternate fuel. The higher its price, the higher the differential value for logging slash. The second important factor is the volume of material salvaged per acre. Other things being equal, the per acre energy value will clearly be higher in old growth Douglas fir stands producing 90 tons per acre than in ponderosa pine stands producing 30 tons per acre. In that case, yarding costs per acre should also be much different.

B. *Smoke Management*

Among the residue removal benefits noted by Kierulff and Adams (1980), perhaps the most important is the fact that slash fires create less smoke after residue is removed. The importance of smoke management has increased over the past decade of increasing environmental awareness. Within the Forest Service, emphasis has shifted from a policy of smoke avoidance to a more active policy of managing emissions at the source. The latter creates an opportunity for residue recovery that serves more than one purpose.

Smoke management by avoidance, forest managers have, in the past, had a great deal of flexibility in the use of prescribed fire as a management tool. Fire has been used to burn logging slash, to release young trees from competing brush, and to reduce fuel loadings that might otherwise foster wildfires. Prescribed fires, because they are generally far from population centers and are of a temporary nature, were not a focus of early pollution control efforts. In 1969, however, Oregon and Washington established a system to reduce the amount of smoke from burning on forest lands. Too much smoke was being carried into or accumulating in certain heavily populated "designated areas" or other areas sensitive to smoke. The State and Federal land management agencies work in cooperation with private forest protection associations and the state environmental agencies to achieve that goal. Cooperators report on their daily burning plans. The states, guided by foresters and meteorologists, determine how many acres can safely be burned and where, in order to stay within national ambient air quality standards (Wash. Dept. Nat. Res., 1982; Sandburg and Schmidt, 1982).

More recently, the State of Washington has proposed stricter air quality standards. Forest burning has come under closer scrutiny. The state has proposed that emissions be reduced by 35% (Wash. Dept. of Ecology, 1983). At the same time, the Department of Ecology is working to expand the Smoke Management Program to include visibility impairment. Visibility standards are being developed for national parks, wilderness areas, and "integral areas" that are outside of, but seen from, national parks and wilderness areas. One way to deal with visibility impairment is to reduce or eliminate burning on weekends during the high use summer season. Although Washington is the only state with this proposed legislation, similar concerns have been expressed in Oregon, California, Arizona, and other states. Similar legislation may follow.

At the same time that stricter air quality limits are being considered, research has been conducted on the cost of the current smoke management by avoidance system. Preliminary results (Sandburg and Schmidt, 1982) indicate that complying with the smoke management program costs the Forest Service $2 million per year in western Washington and Oregon alone. Apportioned over the 76,345 acres burned in an average year, the cost is $26.18 per acre. Of that total, $14.02 is due to changes in operating procedures. Table IV summarizes their findings.

Several changes in operating procedures add to the cost of the smoke management program. The first is work plan changes. The forests are already constrained by the number of suitable burning days available to accomplish their burning goals. If the number of days is reduced in a given year, too many acres must be burned each remaining day. Inefficiencies are introduced because people must be paid overtime, more workers are

TABLE IV. Forest Service Smoke Management Costs in Western Washington and Western Oregon[a]

Cost element	Annual cost	Cost per acre
Total operation changes	$1,070,447	$14.02
Work plan changes	676,211	8.86
Delays	195,237	2.56
Extra work	163,925	2.15
Monitoring and evaluation	35,074	0.46
Treatment changes	625,600	8.19
Administration	303,113	3.97
Total	$1,999,160	$26.18

[a]Source: Sandburg and Schmidt, 1982.

hired for the season, or employees are recruited from other districts to augment the staff temporarily. The additional units burned cost one and one-half to two times as much as comparable units burned. Other work plan changes are made to avoid concentrated smoke in one area. For example, units in a broad geographical area are burned on the same day rather than burning all the units in a single area. The average cost of these adjustments is $8.86 per acre burned.

Other costs are brought about by delays. Smoke management decisions are made on a daily, even hourly, basis so that burning crews might be in the field before a delay decision is made. Sandburg and Schmidt note that suitable burning days are so scarce that about one-fifth of the planned burns delayed one day are not accomplished until the following year. Preparatory work must sometimes be repeated, adding to costs. This amounts to an average of $2.56 per acre burned.

Additional costs are also incurred when areas are burned under drier than optimal conditions. Extra personnel are needed in the field to prevent fire escape and to do mop-up. This extra work adds $2.15 per acre to the cost of smoke avoidance. Other associated costs of the smoke management program include administrative costs and the selection of alternative, more costly fuel treatments.

Given that smoke management by avoidance is not free, and that there is a likelihood that burning days will be further restricted in the future, more attention is being given to another form of smoke management, reducing emissions at the source.

Reducing emissions at the source, Sandburg (1983) has described forest fuelbeds as a mixture of components, including recently deposited foliage, or litter; partially decomposed material, or duff; fine live fuels, including grass and light brush; coarse live fuels, including older brush or undesirable trees; small woody fuels; and large, woody fuels. He concluded that the most attractive technique for reducing emissions from logging slash burning is the increased utilization of large residues.

The gain in emission reduction due to the utilization of the large woody fuels is due to the nature of combustion. Emissions from a given fire are a function of the fuel consumption rate and an emission factor appropriate to the fuel type and fire behavior. The combustion process can be divided into a smoldering phase and a flaming phase, each with a distinct emission factor. Ward (1983) concluded that in complex fuel beds such as logging slash, emission factors are 2 to 2 1/2 times as great during the smoldering phase, which accounts for 1/2 to 2/3 of the consumption and 3/4 of the emissions. Duff is usually too moist to burn alone, so it burns only in conjunction with surface fuels such as logging slash. Removing residues reduces consumption of duff by a like amount, so a twofold reduction in fuel consumption occurs. Emissions are reduced by reducing the smoldering stage of the fire, when most duff and large, woody fuels are consumed (Sandburg, 1980; Little, Ward, and Sandburg 1982, summarized in Sandburg, 1983).

Sandburg and Ward (1982) have reported the results of empirical tests conducted on the same Green Mountain timber sale discussed earlier. Units 2 and 3 were used for the emissions study. Units 2 and 3 were similar in all respects except in the amount of logging slash removed. Unit 2 was YUM yarded to a minimum piece size of 8" × 10'. Logging slash on unit 3 was yarded to a minimum 6" × 6' piece size; pieces larger than 16" × 2' were also removed. Results of the smoke reduction study indicate that particulate emissions were reduced by 660 pounds per acre, or 30% between Units 2 and 3. This was probably the maximum that could be achieved by imposing the 6" × 6' YUM standard on this site. Other sites might have a greater or lesser reduction in smoke. No estimate was made of the difference between a no-YUM burning situation and the 8" × 10' standard.

Given these estimates, we can begin to estimate the per acre value of changing from a smoke avoidance strategy to one which is designed to control emissions at the source. Administrative costs are not expected to change, because they are not directly related to the number of acres treated. The cost of other slash treatments is also not relevant to this dicussion. One can assume, however, that the cost of operational changes can be reduced or eliminated on areas where emissions are controlled at the source. Because of the reduction in emissions, the shorter time required to burn, and the generally improved efficiency of the burn, the problems associated with smoke avoidance are mitigated. Some problems will still exist in areas that must be burned. Class 1 airsheds such as National Parks will still have to be considered in burn plans, for example. We assume, therefore, that yarding to smaller specifications is worth from zero to $14.02 per acre in reduced smoke management costs. That savings is shown on Table III.

It is more difficult to determine the value of the reduction in smoke itself. What is 660 pounds of smoke worth exactly? Given the inherent problems in dealing with this question, the implied value of reduced emissions will be derived using the procedure presented in Section V. All other costs and benefits will be determined, and the level of emission control will be compared to the remaining net present value.

C. Soil Disturbance

There is an element of controversy surrounding soil disturbance and nutrient loss due to residue removal. Braunstein and others (1981) have summarized potential impacts of biomass harvesting on soils and nutrients. The effects of residue removal fall into three categories: removal of the biomass itself, soil compaction, and disturbance of the litter layer. Removal of the biomass itself creates an immediate loss of organic matter. Soil compaction, especially on gentle terrain where tractors are used, decreases the infiltration of precipitation, which leads to increased runoff, accelerated erosion, and a continuing loss of soil nutrients. Disturbance of the litter layer can have the same effects.

On the other hand, prescribed fire volatizes nitrogen and sulfur in residue and duff and alters the chemical nature of the soil (Wells et al., 1979). Chemical alteration can lead to increased loss of nutrients through leaching. Boyer and Dell (1980) have established maximum acceptable limits for duff consumption and area of mineral soil exposed. The limits are designed to protect both water quality and the nutrient capital of the soil. Results on the Green Mountain timber sale have been summarized by Little, Ward and Sandburg (1982). Both units were within the maximum limits set by Boyer and Dell. On unit 3, the reduced fuel load associated with a 6" × 6' slash removal standard corresponded to a 35% reduction in fine fuel consumption and a 36% reduction in duff consumption. The proportion of mineral soil exposed was also reduced. These results indicate that additional slash removal can have a positive effect on the retention of nutrient capital when compared to burning under normal removal standards. Given the uncertainty surrounding the question of soil damage, this analysis assumes that the impact is negligible. It will be treated neither as a cost nor a benefit of yarding unmerchantable logging slash. More research is needed to determine the impact of intensive yarding on soils.

D. Improved Ground Access and Earlier Reforestation

As mentioned earlier, one of the primary reasons for prescribed burning is to prepare a site for planting. Burning over clearcut areas has therefore met the dual purpose of slash disposal and site preparation. Increasing restrictions on the number of burning days available means an increased likelihood that acres harvested in one year will not be prepared for regeneration until at least the following year. Delay in site preparation and planting gives competing brush a chance to take hold, compounding later regeneration problems.

How much is it worth to gain that extra year for a new stand of timber? Figures developed for the Willamette National Forest land management plan can provide a useful answer to that question (Ullrich, 1983). Table V summarizes the cost of stand establishment and management, plus the costs and benefits of final harvest. All costs and benefits are discounted to their present value in the year the existing stand is harvested, using a real discount rate of 3%. Harvest of the regenerated stand is assumed to be at age 85, which roughly equals the age of maximum economic value estimated by the Forest's land management plan. Case 1 assumes that the site is prepared and planting takes place in year zero, the year of harvest. Case 2 assumes that site preparation and planting are delayed one year. The value of gaining one year in stand establishment is equal to the difference between the net present values for Cases 1 and 2.

As indicated by Table V, the value of planting one year earlier is positive, but not substantial. The results are necessarily sensitive to the specific assumptions regarding costs, particularly planting costs. It is possible that the planting cost in Case 2 should be higher to reflect the additional access problems associated with higher slash loadings and one year's growth of brush. The results are also sensitive to the assumption made about real escalation in the price of stumpage. If one assumes that there is no real price

TABLE V. Net Present Value of a Regenerated Stand[a]

		Case 1		Case 2	
Activity	Cost/ acre	Year	Present value	Year	Present value
Plant, replant	499	0	- 499	1	- 484
Release (20% of ac. @ $83/acre)	17	8	- 13	9	- 10
Precommercial thin	142	12	- 100	13	- 97
Sale prep/admin ($6.37/mbf and 50 mbf/acre)	318	85	- 26	86	- 25
Stumpage value @$331/mbf	16550	85	1342	86	1303
Net present value			704		687

Difference in value per acre = $17

[a] Source: Calculated from data provided by Willamette National Forest (Ullrich, 1983).

escalation, that is, that the prices bid for stumpage change at the same rate as inflation, then the additional value of planting in the year of final harvest is only $17 per acre. If one were to assume that stumpage prices will rise faster than the rate of inflation, such as they have in all but the recent past, then the value of planting a year early would increase. The value of $17 per acre is included on Table III as the value of earlier reforestation.

E. Reduced Risk of Wildfire

The reduced risk of wildfire comes in two forms. First, reduced fuel loadings on intensively YUMed areas should reduce the probability of wildfires starting in the area. Secondly, the reduced duration of the prescribed burn, particularly the reduction in the smoldering phase, should reduce the incidence of escaped slash fires. This second situation is the more significant. Slash fires tend to escape their prescribed boundaries when fresh winds stir up smoldering material. More than

75% of the largest fires (100 acres or more) are the result of escaped slash fires (Nor'west Pacific Corp., 1981). The Forest Service estimated cost of suppression of these larger fires is approximately $500,000 per fire. In addition, about 70% of smaller fires are also caused by escaped slash fires. These are frequently less than one-quarter acre in size.

The actual cost of suppressing escaped slash fires is not known, nor is it entirely clear how many fires could be avoided by more intensive YUMing. It is apparent, however, that the increased practice of YUM over the last ten or fifteen years, plus the use of improved technology and altered burning procedures, has already reduced the incidence of escaped fires. For the purposes of this analysis, assume that the widespread use of the 8" × 10' YUM standard reduces wildfire suppression costs and timber losses in western Washington and Oregon by $1 million dollars per year, and that the 6" × 6' standard would reduce costs by another $1 million. Dividing by the average number of acres burned per year produces an estimate of the average benefit per acre.

8" × 10' units: $1,000,000 - 76,345 acres = $13.10/acre

6" × 6' units: $2,000,000 - 76,345 acres = $26.20/acre

This assumption provides only a rough estimate of the benefits of wildfire reduction. Further research is needed to determine the actual expected benefits. These estimates are shown on Table III as the maximum benefit of reducing wildfire risk.

F. Reduced Burning Cost

As mentioned earlier, slash fires in intensively YUMed areas have a shorter smoldering phase than other slash fires. The reduced length of this phase reduces the time needed for mop-up, and therefore reduces the cost of mop-up. It is generally believed that mop-up costs on a $250-350 per acre prescribed burn can be reduced by about $50 per acre when more material is removed (Mapes, 1983). That savings is shown on Table III as a reduced cost of burning.

G. Reduced Planting Cost

More research is needed to determine the extent to which residue removal reduces planting costs. It is reasonable to assume, however, that some savings are enjoyed. Removal of the material eliminates obstacles and increases the potential productivity of the planting crew. This analysis assumes the cost savings to be $50 per acre, as shown on Table III. Because planting costs on other forests are generally lower than those on the Willamette National Forest, the savings per acre might also be much lower elsewhere.

H. Results of the Analysis

The results of these estimates are shown on Table III. The cost absorbed by the Forest Service through lower stumpage values are $1496 and $2484 respectively for the less intense and more intense YUMing. The value of all benefits for which prices could be estimated reached maximums of $1037 and $1372 respectively. The remaining net present values were therefore negative for both standards. That is, the costs exceed the value of the priced benefits.

To determine whether a policy for YUM yarding is still in the public interest, the net present value is compared to the amount of smoke reduced on each acre. Reduction on the 8" × 10' units is assumed to be the average estimated on experimental sites (Sandberg and Ward, 1982). The additional reduction of 660 lbs. per acre, for a total reduction of 1024 lbs. per acre on the 6" × 6' units, is consistent with the experience on the Green Mountain sale. By dividing the number of pounds of smoke reduced into the net present value of each unit, the implied net cost of the smoke reduction is calculated. That is, we calculate the amount that the smoke would have to be worth in order to make either YUM standard economically efficient.

Given the values estimated for benefits of YUMing to an 8" × 10' standard, the implied cost per pound of smoke reduced is $1.26 to $1.30. On the 6" × 6' standard units, the cost would be only $1.09 to $1.14. Do these costs seem reasonable? Emission control in other industries such as concrete, lime, asphalt, iron and steel generally costs about $200 to $1000 per ton of reduced emissions. That equals $0.10 to $0.50 per pound, less than the estimates made in this example of slash burning.

The citizens of the United States have decided, by their choice to implement the Clean Air Act, that air pollution control is worth at least $0.10 to $0.50 per pound of smoke reduced. Although most individuals did not make such a decision explicitly, it is implied by the fact that the law exists and is enforced. Other sites can be evaluated using the same procedure. The closer the resulting estimate comes to the $0.10 to $0.50 range per pound of smoke reduced, the more appropriately one can conclude that YUMing to either standard maximizes net public benefit.

As stated earlier, this analysis is not meant to produce a definitive answer to the question "Does YUM yarding maximize net social benefit?" It does, however, present the framework within which the question can be approached. The following section presents some other aspects to the analysis. First, opportunities to reduce YUM costs are discussed. There follows a discussion of a different management approach to determining when and where YUM should be required.

VII. REDUCING THE COST OF YUM YARDING

The costs of YUM yarding on the Green Mountain sale are not necessarily representative of the average cost to YUM. Hanscom (1979) cites experienced costs in Oregon of $387 to $1450 per acre in 1979 dollars ($530 to $1986 in 1982 dollars). Similarly, a cost of approximately $800 to $900 is not unusual in north Idaho forests (Crosser, 1983). The cost of YUMing is influenced by the terrain, the equipment used, size of the material being moved, the skill, and even the morale, of the crew. Hanscom (1979) suggests that "everyone should try to set an inch or an inch and one-eighth choker on a 9" madrone and get it to the landing."

The equipment used has an important impact on the cost of yarding unmerchantable material. The small unmerchantable material is frequently yarded with the same equipment used for the merchantable logs. The powerful and expensive equipment required for cable logging on steep slopes is best used to remove large timber. Most systems capable of handling large material are designed for that purpose, so that removal of small material reflects an inefficient use of the machinery. There is also a large opportunity cost associated with using large machinery to yard unmerchantable wood when it could have been in use recovering higher valued material (Bergvall and others, 1978).

Even given these inefficiencies, it is sometimes considered cheaper in the long run to yard all material with the same equipment. This will at least save the cost of having two kinds of equipment available, and transporting and setting up two systems on every unit. New equipment is always being developed, however, which can reduce the cost of reentry to yard unmerchantable material.

Brown and Bergvall (1983), for example, report on the use of a Bitterroot Miniyarder to remove residue from previously harvested lands managed by the Washington Department of Natural Resources. They estimate the falling, bucking, yarding and decking of fuelwood in the study cost $15.66 per green ton, not counting overhead. This cost is still fairly high, especially given the fact that inexpensive labor from a Washington Corrections Center was used. The net cost of using the miniyarder was lower, however, than the alternative methods available for rehabilitating the same sites.

In another study based on the Green Mountain timber sale data, Brown (1982) performed a double entry yarding simulation to see if yarding costs could be reduced. Double entry yarding assumes that only large, heavy logs would be yarded with the large machinery. When the heavy material has been removed, a smaller yarder would be brought in to yard the remaining small material. Unit 1A of the Green Mountain sale was used for the analysis. The unit was yarded to the smaller 6" × 6' specification.

The actual cost of yarding unit 1A was $76,688 for the 16.5 acre unit. Brown's simulation, assuming the use of a Mini-Alp yarder for the second entry, reduced the total cost to $50,083. The simulated first entry cost $42,134 to yard the merchantable material. The second entry estimated removal of the smaller unmerchantable material at a total cost of $7949. Yarding cost per ton was reduced from $13.40 to $8.75, a reduction of 35%.

To test the sensitivity of the earlier analysis to lower yarding costs, the net present value of the residue removal can be recalculated using the lower cost. Table VI summarizes this analysis. It is identical to Table III except that the yarding cost has been reduced by 35%.

The results of this analysis indicate that, for this example, reducing yarding costs by 35% results in a 53 to 73% reduction in the implied cost of smoke reduction. The cost of $0.34 to $0.54 per pound is within the cost range for reducing other forms of emissions. It is therefore more reasonable to assume that in this case, net public benefit would be maximized if the Forest Service absorbs the cost of providing wood for the fuel market. While this example is based only on simulated yarding costs, it illustrates the effect that reduced costs can have on the overall analysis.

VIII. SITE SPECIFIC DECISION MAKING

A cost-benefit analysis such as the one presented here could indicate whether it would be worthwhile, on a specific sale, to include in the timber sale appraisal the costs to yard, load, and/or haul unmerchantable material. It cannot be taken for granted that YUMing is beneficial because it provides wood for energy and reduces air pollution. Nor can it be taken for granted that YUMing is not worthwhile because it is so expensive. Each case must be considered individually.

TABLE VI. *The Implied Value of Smoke Standard and Intensive YUMMING Compared to No YUM Simulated Double Entry Yarding Costs Assumed*

	Standard YUM	Intensive YUM
	>8"×10'	>6"×6'
	(Dollars per acre)	
Costs absorbed by Forest Service		
Yard	621	1136
Load	328	466
Haul to disposal site	213	270
Total Cost	1162	1872
Benefits		
Net energy value (equals value/ton delivered to mill less haul cost/ton from disposal site to mill times tons/acre)	924	1215
Smoke management cost avoided	0	0 to 14.02
Present value of earlier reforestation	0	0 to 17.00
Reduced risk of wildfire	0 to $13.10	0 to 26.20
Reduced burning cost	50	50
Reduced planting cost	50	50
Total priced benefits	1024 to 1037	1315 to 1372
Net present value	-138 to -125	-557 to -500
Physical smoke reduction	364 lbs.	1024 lbs.
Implied value per pound of smoke reduction	0.34 to 0.38	0.49 to 0.54

Public land managers are responsible for the efficient management of all the resources at their disposal—land, labor, and capital. Their goal is to maximize net public benefits, and that goal is accomplished through the efficient use of resources. In an earlier era, maximizing net public benefits was often seen as synonymous with maximizing returns to the U.S. Treasury. Certainly that is to the advantage of all taxpayers. But wise land use also requires consideration of those unpriced resources which society also values, such as clean air and water. The tradeoff between cash income and unpriced benefits should always be made as explicit as possible, and decisions made accordingly.

A case in point is the Washington Department of National Resource's management guidelines. The department manages 1,833,000 acres of forest land in Washington. The department retains 25 to 50% of its gross receipts from land management activities such as timber sales. This revenue is used to finance all its activities, including all salaries and administrative expenses. The remainder of the receipts are paid into eleven trusts which own the land managed by the Department. The trusts provide support to several schools, counties, and state charitable and penal institutions.

Because the department's operating budget is directly dependent upon its gross receipts, and because the trusts depend on the income from timber sales, there is an incentive to maximize receipts. The department's philosophy is more balanced, however. It has "a dedication to produce revenues from the forests to continue support of public schools and other institutions, but also a commitment to diversity investments and practices to assure a continuance of those revenues for the public, for whom the lands are a trust." At times this means the sacrifice of current revenue to produce other, nonpriced, benefits.

The Proposed Forest Land Management Program (Wash. Dept. Nat. Res. 1982, p. 148) states that the department "will make wood fiber available for energy whenever it cannot be used as a higher valued product or when costs of supplying it do not exceed benefits." The department's consideration of benefits specifically includes smoke management, wildfire risk, water quality effects, and soil nutrient levels. The department will use its evaluation of these upriced benefits to justify some projects, "even though this may result in a fiber energy project that cannot stand on its own economically."

The success of such a philosophy depends on careful evaluations made on a site specific, project specific basis.

IX. SUMMARY

The intent of this chapter was to outline a procedure for evaluating the benefits and costs of recovering logging, residue for energy use. The principal conclusions are as follows:

(1) The demand for wood as an energy source is derived from the demand for final products: electricity, process steam for drying plywood, and so on. The current price of electricity in the Pacific Northwest limits the price that can be paid for fuelwood to generate electricity.

(2) The price of alternative fuels also places a limit on the price that potential customers can pay for logging residue. Logging residue cannot be sold for $20 per ton if mill residues are readily available at $15 per ton.

(3) The cost of yarding unmerchantable material (YUM) is the major factor in determining the overall cost of providing logging residue for fuel. Cost estimates vary widely, depending on the site.

(4) The price at which logging residue can be profitably sold is lower if the Forest Service pays for its removal by accepting a lower price for merchantable timber on the sale.

(5) There is a method to determine if it is worthwhile for the Forest Service to pay for the removal of wood for energy. The procedure follows a 4-step process:

(a) Determine all relevant costs and benefits. Prices and costs are estimated for as many factors as possible.
(b) Calculate the net present value of each management alternative, that is, determine the value of benefits minus cost.
(c) If the costs exceed the benefits, compare the net cost to the remaining unpriced benefits.
(d) Make a subjective judgment about whether the remaining benefits are worth the cost of acquiring them. Where it is feasible to do so, compare the cost of acquiring these benefits with the cost of acquiring similar benefits through other means.

(6) More research is needed to provide better estimates of the dollar value of benefits. Specific areas needing more research include the effects of intensive yarding on soils, the effect on wildfire risk, and the effect on subsequent management costs.

(7) Following this analytical procedure on the Green Mountain sale indicated that the cost of smoke reduction was

somewhat higher than that found in other industries. The evaluation of other sites could produce different results.

(8) The cost of emission control on the Green Mountain sale could be reduced if yarding costs could be reduced by the use of a miniyarder for the smaller material.

(9) Results would be even better if the removal of logging residue could completely eliminate the need to burn harvested areas in order to dispose of the remaining slash and prepare the site for planting.

(10) The real answer to the question of whether or not residue removal is worthwhile can only be determined on a site specific, project specific basis. The Forest Service should have the flexibility to make those determinations.

X. NOTES

(1) the logging contractor might not be indifferent to the amount of work required if there are more profitable uses for the same equipment and labor; for example, yarding merchantable logs on a new sale area. Bergvall and others (1978) have noted the large opportunity cost associated with the inefficient use of machinery to yard unmerchantable wood when it could have been employed on higher valued products.

(2) For a thorough discussion of the rationale behind the sustained yield policy in general and the nondeclining evenflow policy in particular, see LeMaster and others (1982).

(3) This discussion assumes that the net value of using logging residue as fuel is captured by the mill owner in the form of lower fuel costs. However, the benefit could be received by the logging contractor (or agent) in the form of higher prices charged for the material he recovers and sells for fuel, or the benefit could be split between buyer and seller. It could also be passed on to consumers as lower end-product prices. The relevant point is not who gets the benefit, but the fact that a net benefit has been created. Assuming that the benefit goes to the mill owner or other residue buyer simplifies the calculation of net energy value.

APPENDIX TABLE A-1

UNITS YARDED TO 8" X 10'

UNIT 1B = 16.5 ACRES

MATERIAL	NET WGT (TONS)	COST PER TON (1980 $$)		
		YARD	LOAD	HAUL TO DISPOSAL SITE(1)
OP	23.9	12.45	3.27	2.00
WF	1208.7	10.18	2.49	2.00
WF-CHIPS	192.8	11.87	8.84	2.00
TOTAL				
TONS	1425.4			
COST-1980$		14891	4792	2851
COST-1982$		17574	5656	3365

UNIT 2 = 16 ACRES

MATERIAL	NET WGT (TONS)	COST PER TON (1980 $$)		
		YARD	LOAD	HAUL TO DISPOSAL SITE(1)
OP	6.5	25.93	10.46	2.00
WF	1502	7.49	2.77	2.00
WF-CHIPS				
TOTAL				
TONS	1508.5			
COST-1980$		11419	4229	3017
COST-1982$		13476	4991	3561

TOTAL ALL 8" X 10'				
TONS	2933.9			
COST-1980$		26309	9021	5868
COST-1982$		31051	10646	6925
TONS PER ACRE	90			
1982$ PER ACRE		955	328	213

APPENDIX TABLE A-1 (CONTINUED)

UNITS YARDED TO 6" X 6'

UNIT 1A = 16.5 ACRES

MATERIAL	NET WGT (TONS)	COST PER TON (1980 $$)		
		YARD	LOAD	HAUL TO DISPOSAL SITE(1)
OP	23.6	13.24	2.67	2.00
WF	1375.6	13.15	3.19	2.00
WF-CHIPS	210.9	19.33	8.19	2.00
TOTAL				
TONS	1610.1			
COST-1980$		22478	6178	3220
COST-1982$		26529	7292	3801

UNIT 3 = 17 ACRES

MATERIAL	NET WGT (TONS)	COST PER TON (1980 $$)		
		YARD	LOAD	HAUL TO DISPOSAL SITE(1)
OP	47.2	7.27	2.27	2.00
WF	2006.2	7.94	2.36	2.00
FIREWOOD	256.5	22.86	8.02	2.00
TOTAL				
TONS	2309.9			
COST-1980$		22136	6899	4620
COST-1982$		26125	8142	5452

UNIT 4, 4A = 39 ACRES

MATERIAL	NET WGT (TONS)	COST PER TON (1980 $$)		
		YARD	LOAD	HAUL TO DISPOSAL SITE(1)
OP	41.6	12.16	3.03	2.00
WF	4778.6	12.27	2.99	2.00
FIREWOOD	290.4	46.42	9.42	2.00
TOTAL				
TONS	5110.6			
COST-1980$		72620	17150	10221
COST-1982$		85707	20240	12063

UNIT 5 = 15 ACRES

MATERIAL	NET WGT (TONS)	COST PER TON (1980 $$)		
		YARD	LOAD	HAUL TO DISPOSAL SITE(1)
OP	45.6	10.68	3.09	2.00
WF	549.3	10.75	3.13	2.00
FIREWOOD	387.8	15.3	6.34	2.00
TOTAL				
TONS	982.7			
COST-1980$		12325	4319	1965
COST-1982$		14547	5097	2320
TOTAL ALL 6" X 6'				
TONS	10013.3			
COST-1980$		129559	34546	20027
COST-1982$		152908	40772	23636
TONS PER ACRE	114			
1982$ PER ACRE		1748	466	270

NOTE (1) - APPRAISAL ALLOWANCE FOR HAUL TO DISPOSAL SITE ASSUMED TO BE $2.00 PER GREEN TON

SOURCE: ADAPTED FROM BROWN (1981). TABLES A-III, A-IV, A-X AND VII

APPENDIX TABLE A-2

DERIVATION OF HAUL COST FROM DISPOSAL SITE TO MILL

UNITS YARDED TO 8" X 10'

UNIT 1B

MATERIAL	NET WGT (TONS)	HAULING COST PER GREEN TON (1980 $$)		
		TOTAL	TO DISP SITE	DISP SITE TO MILL
OP	23.9	5.14	2.00	3.14
WF	1208.7	5.35	2.00	3.35
WF-CHIPS	192.8	10.00	2.00	8.00
TOTAL				
TONS	1425.4			
COST-1980$		8517	2851	5667
COST-1982$		10052	3365	6688

UNIT 2

MATERIAL	NET WGT (TONS)	HAULING COST PER GREEN TON (1980 $$)		
		TOTAL	TO DISP SITE	DISP SITE TO MILL
OP	6.5	26.18	2.00	24.18
WF	1502	5.95	2.00	3.95
WF-CHIPS				
TOTAL				
TONS	1508.5			
COST-1980$		9107	3017	6090
COST-1982$		10748	3561	7188

TOTAL ALL 8" X 107				
TONS	2933.9			
COST-1980$		17624	5868	11757
COST-1982$		20801	6925	13875
1982$ PER TON		7.09	2.36	4.73

UNITS YARDED TO 6" X 6'

UNIT 1A

MATERIAL	NET WGT (TONS)	HAULING COST PER GREEN TON (1980 $$)		
		TOTAL	TO DISP SITE	DISP SITE TO MILL
OP	23.6	5.08	2.00	3.08
WF	1375.6	5.35	2.00	3.35
WF-CHIPS	210.9	10.00	2.00	8.00
TOTAL				
TONS	1610.1			
COST-1980$		9588	3220	6368
COST-1982$		11316	3801	7516

UNIT 3

MATERIAL	NET WGT (TONS)	HAULING COST PER GREEN TON (1980 $$)		
		TOTAL	TO DISP SITE	DISP SITE TO MILL
OP	47.2	5.13	2.00	3.13
WF	2006.2	5.20	2.00	3.20
FIREWOOD	256.5	5.46	2.00	3.46
TOTAL				
TONS	2309.9			
COST-1980$		12075	4620	7455
COST-1982$		14251	5452	8799

UNIT 4, 4A

MATERIAL	NET WGT (TONS)	HAULING COST PER GREEN TON (1980 $$)		
		TOTAL	TO DISP SITE	DISP SITE TO MILL
OP	41.6	5.79	2.00	3.79
WF	4778.6	5.43	2.00	3.43
FIREWOOD	290.4	12.26	2.00	10.26
TOTAL				
TONS	5110.6			
COST-1980$		29749	10221	19528
COST-1982$		35110	12063	23047

UNIT 5

MATERIAL	NET WGT (TONS)	HAULING COST PER GREEN TON (1980 $$)		
		TOTAL	TO DISP SITE	DISP SITE TO MILL
OP	45.6	5.28	2.00	3.28
WF	549.3	5.45	2.00	3.45
WF-CHIPS	387.8	5.70	2.00	3.70
TOTAL				
TONS	982.7			
COST-1980$		5445	1965	3480
COST-1982$		6426	2320	4107
TOTAL ALL 6" X 6'				
TONS	10013.3			
COST-1980$		56857	20027	36830
COST-1982$		67104	23636	43468
1982$ PER TON		6.70	2.36	4.34

SOURCE: ADAPTED FROM BROWN (1981), TABLES VII, A-III, A-IV, & VII

REFERENCES

Anderson, Steve. 1983. Washington Water Power Co., Spokane Washington. Personal communication, September, 1983.

Benson, Robert. 1983. USDA Forest Service Intermountain Forest and Range Experiment Station, Missoula, Montana. Personal communication, August, 1983.

Bergvall, John A., and Brown, Steven L. 1983. Washington Department of Natural Resources, Olympia, Washington. Personal communication, August, 1983.

Bergvall, John A., Bullington, Darryl C., Gee, Loren, and Koss, William. 1978. Wood Waste for Energy Study: Inventory Assessment and Economic Analysis. Washington Department of Natural Resources, Sept. 1, 1978.

Bioenergy Bulletin. 1982. Idaho Utility Buys Power from Sawmill Cogenerator. Vol. 3, No. 5.

Bonneville Power Administration. 1983. Pacific Northwest and Alaska Bioenergy Program Annual Report, February, 1983.

Boyer, Donald E., and Dell, John D. 1980. Fire Effects on Pacific Northwest Soils. R6-WM-040-1980. USDA Forest Service Pacific Northwest Region, Portland, Oregon.

Braunstein, Helen M. 1981. "Biomass Energy Systems and the Environment." Oak Ridge National Laboratory, Pergamon Press, Elmsford, New York.

Brown, Larry and Associates, Inc. 1981. Green Mountain Timber Sale: Wood Residue Utilization Study. USDA Forest Service Willamette National Forest, July, 1981.

Brown, Larry and Associates, Inc. 1982. Further Investigation into Constraints and Economics of Residue Collection. USDE Bonneville Power Administration, August, 1982.

Brown, Steve L., and Bergvall, John A. 1983. In-woods Testing and Feasibility Study of Fuelwood Recovery Using a Small Scale Cable Yarding System. Washington State Energy Office, July, 1983.

Crosser, A. 1983. USDA Forest Service Idaho Panhandle National Forest. Personal communication, April, 1983.

Dorn, Karen. 1983. California Wants That Power Surplus. The Spokesman Review, June 26, 1983.

Federal Register. 1979. National Forest System Land and Resource Management Planning. 44:53928-53999.

Geomet, Inc. 1978. Impact of Forestry Burning Upon Air Quality, Final Report. EPA 910-9-78-052. U.S. Environmental Protection Agency, Region X, Seattle, Washington.

Hanscom, Ed. 1979. YUM Costs - Is it Worth it? *Forest Industries,* August, 1979.

Howard, James O. 1979. Wood for Energy in the Pacific Northwest: An Overview. USDA Forest Service, Pacific Northwest Forest and Range Experiment Station Gen. Tech Rep. PNW-94.

_____. 1981a. Logging Residue in the Pacific Northwest: Characteristics Affecting Utilization. USDA Forest Service, Pacific Northwest Forest and Range Experiment Station Res. Paper PNW-289.

_____. 1981b. Ratios for Estimating Logging Residue in the Pacific Northwest. USDA Forest Service, Pacific Northwest Forest and Range Experiment Station Res. Pap. PNW-288.

Hyde, William F. 1980. Timber Supply, Land Allocation and Economic Efficiency. Resources for the Future, Johns Hopkins Univ. Press, Baltimore, Maryland.

Johnson, A. R. 1982. Endurance Firing of Wood-Coal Combination at Washington Correction Center, Shelton, Washington. Quinault-Pacific Corporation.

Juhasz, Paul. Washington State Energy Office. Personal communication, August, 1983.

Keegan, Charles E. 1981. The Cost and Availability of Forest Residue in the Northern Rocky Mountains. Bureau of Business and Economic Research, University of Montana, Missoula, Montana.

Kierulff, Neil C., and Adams, Thomas C. 1980. Feasibility of Generating Electric Power from Forest Residue. USDA Bonneville Power Admin. and USDA For. Serv. Pacific Northwest Forest and Range Experiment Station.

Krutilla, John V., and Fisher, Anthony C. 1975. The Economics of Natural Environments. Resources for the Future, Johns Hopkins Univ. Press, Baltimore, Maryland.

LeMaster, Dennis C., Baumgartner, David M., and Adams, David, eds. 1982. Sustained Yield: Proceedings of a Symposium held April 27 and 28, 1982, Spokane, Washington. Washington State Univ. Coop. Extension.

Little, Susan N., Ward, Franklin R., and Sandburg, D. V. 1982. Duff Reduction Caused by Prescribed Fire on Areas Logged to Different Management Intensities. USDA Forest Service, Pacific Northwest Forest and Range Experiment Station Res. Note PNW-397, May.

Mapes, Hubert. 1983. USDA Forest Service Willamette National Forest. Personal communication, August, 1983.

Martin, R. E. 1978. Prescribed Burnings: Decisions, Prescription, Strategies. Presented at the 5th National Conference on Fire and Forest Meteorology, American Meteorological Society and Society of American Foresters.

Morgan, Richard E. 1983. Supreme Court Saves Small Power Rules. *Power Line,* June, 1983.

Northwest Power Planning Council. 1983. Northwest Conservation and Electric Power Plan, Vol. 1. Portland, Oregon.

Nor'west Pacific Corporation. 1981. Feasibility Study for a Forest-Residue-Fueled Electric Generating Plant. Technical Planning Study TPS 79-742. Prepared for Electric Power Research Institute and Eugene Water and Electric Board, Eugene, Oregon.

Pacific Northwest Utilities Conference Committee. 1982. Wood Residue Energy for Utilities.

Pacific Power and Light. 1982. Avoided Cost Prices for Purchased Power. Portland, Oregon, November 30, 1982.

Puget Sound Power and Light Company. 1983. Avoided Energy Cost. PSP&L, Bellevue, Washington. Revised May 31, 1983.

Rocket Research Company. 1980. A Study of the Feasibility of Cogeneration Using Wood Waste as Fuel. Technical Planning Study TPS 79-736-1. Prepared for Electric Power Research Institute and Lewis County Public Utility District No. 1, Lewis County, Washington.

Sandburg, D. V. 1983a. Emission Reduction for Prescribed Burning. USDA Forest Service, Pacific Northwest Forest and Range Experiment Station. Presentation to the 76th Annual Meeting, Air Pollution Control Association, Atlanta, Georgia, June 19-24.

_____, 1983b. USDA Forest Service Pacific Northwest Forest and Range Experiment Station, Seattle, Washington. Personal communication, August, 1983.

_____, and Schmidt, R. Gordon. 1982. Smoke Management Costs for Forest Burning. USDA Forest Service Pacific Northwest Forest and Range Experiment Station. Presentation to Air Pollution Control Association Pacific Northwest International Section Annual Meeting, Vancouver, B.C., Nov. 15-17, 1982.

_____, and Ward, D. E. 1982. Increased Wood Utilization Reduced Emissions from Prescribed Burning. USDA Forest Service Pacific Northwest Forest and Range Experiment Station. Presentation to the 1982 West Coast Regional Meeting, National Council of the Paper Industry for Air and Stream Improvement, Inc., May 18, 1982.

Ullrich, Richard. USDA Forest Service Willamette National Forest. Personal communication, September, 1983.

USDA Forest Service, 1977. Forest Service Manual FSM 2403.25, Timber Management; Utilization.

Washington Department of Ecology. 1983. Revision to the Washington State Implementation Plan, Washington State's Visibility Protection Program. Washington Department of Ecology, Division of Air Programs, February, 1983. Olympia, Washington.

Washington Department of Natural Resources. 1982a. Draft Enviromental Impact Statement: Forest Land Management Program. Washington Department of Natural Resources, Olympia, Washington.

_____. 1982b. Proposed Forest Land Management Program. Washington Department of Natural Resources, November, 1982. Olympia, Washington.

Washington Water Power Company. 1983. Schedule 62: Small Power Production and Cogeneration Schedule, Washington. First revision dated May 24, 1983. The Washington Water Power Company, Spokane, Washington.

REVIEW OF BIOMASS GASIFICATION TECHNOLOGY

Kenneth L. Tuttle

U. S. Naval Academy
Annapolis, Maryland

I.	INTRODUCTION	264
II.	SOURCES OF INFORMATION ON BIOMASS GASIFICATION	264
	A. Marenco 1982	265
	B. Miller 1983	265
	C. Oliver 1982	266
	D. Reed 1979	266
	E. Tuttle 1978	267
III.	GASIFICATION PROCESSES	267
	A. Fluid Bed Gasifier	267
	B. Down Draft Gasifier	269
	C. Fixed Bed Updraft Gasifier	270
IV.	COMMERCIAL AVAILABILITY OF GASIFICATION	277
	A. Units Operating Commercially	277
	B. Other Experienced Manufacturers	278
V.	SUMMARY	278

INTRODUCTION

Biomass gasification is a diverse collection of technologies, and it is becoming more diverse every year. Within the past three years two processes have been successfully implemented. Without intimate knowledge of installation, it is often difficult to tell the successes from the near successes. A review of the state-of-the-art in biomass gasification may aid the discerning observer.

The information available on gasification is large and unwieldy. This approach is to review the best sources of in-depth information available to the author. The references listed give detailed descriptions of the three currently commercial biomass gasification processes. They are supplemented with brief descriptions and comparisons of the processes. This review identifies those commercially operating biomass gasifiers known to be meeting design specifications successfully. Also it identifies manufacturers whose commercial gasifiers are in startup.

Biomass gasification has broken through several barriers in the past few years. The successes are the results of commendable team efforts. The problems associated with these systems were by no means limited to engineering. As with the emergence of any new technology, the number of obstacles to be overcome was discouragingly large. Furthermore, none of the gasification development teams had large budgets. Yet the technical and economic feasibility of biomass gasification is now being demonstrated in several locations by commercially operating gasifiers.

II. SOURCES OF INFORMATION ON BIOMASS GASIFICATION

The five references listed are good sources of information on the three gasification processes which are being used commercially for biomass. Between them, they cover the principles of gasification, current technology and research, and available manufacturers. Substantial information and data from testing are reported in the references. Unfortunately, most of the testing which has been conducted on operating biomass gasifiers is not reported in the references and may not exist in the public domain.

Testing of downdraft gasifiers was reported extensively (Marenco, 1982) and by the University of California, Davis. The fluid bed gasification process has been tested nearly as extensively but reported sparingly (Miller, 1982, and Oliver, 1982). The fixed bed process, which produces relatively cool gas loaded with condensed hydrocarbons, is operating and has been tested. The few data available publicly are from pilot plant operations (Tuttle, 1978) and do not include tars and oils.

Historical data and recent prior data reported for the fixed bed process are out of date because the processes which produced them do not approximate closely the fixed bed gasifier which is operating commercially.

A. *Marenco 1982*

Marenco in its final report in March 1982 reported on a "Wood Gasification/Power Generation Development Project" for remote Alaskan villages. The Alaskan program concentrated effort on the development of a simple wood gasification power generation system which would be suitable for use in rural Alaskan communities. In such remote villages, the power requirements are low. However, the price of fuel oil is exceptionally high and in some cases wood fuel can easily be transported from nearby sources.

The program objective was to demonstrate a simple, reliable wood gasification/power generation system. All components of the system were already available and all aspects of using the system were demonstrated. The major pieces of equipment were a three million Btu per hour downdraft gasifier designed and built by Biomass Corporation and a Caterpillar G353 spark ignition engine rated at 225KW of natural gas.

The Institute of Gas Technology reviewed the work. The system was not an unqualified success. There are more problems firing gasified biomass in an engine than in a burner. Substantial progress was made however, and we gained valuable knowledge.

B. *Miller 1983*

Miller in the final report on "State-of-the-Art Survey of Wood Gasification Technology" provided a summary of the technology available to produce low Btu gas to displace gas or oil in small boilers. The report touches on wood resource availability and economics, in addition to surveying both domestic and foreign manufacturers of wood gasifiers. As of December 1982, thirty-five manufacturers were reported. Most manufacturers were developing a commercial unit or still studying pilot plants. Five had no pilot plant, and five had commercial units in operation. The report provides location as well as detailed information on each manufacturer. The data provided by the report will also be of great interest. The fluid bed gasifier produced by Omnifuel is of particular note.

C. *Oliver 1982*

Oliver's final report on a "Technical Evaluation of Wood Gasification" provides documented performance information of commercial biomass gasifiers. Installed commercial gasifiers were assessed for operability and performance. Only the Ominfuel fluidized bed gasifier at Hearst, Ontario, met the criteria selected. The data provided a reasonable indication of the gasifier's performance in spite of gaps and inconsistencies. No long term data were available at that time on operation and maintenance. The gasifier was observed to be responsive to its controls and to give acceptable performance.

The fixed bed gasifiers manufactured by Forest Fuels and Applied Engineering Company were not observed closely enough to permit accurate information to be reported. However, the reader is introduced to these two successful manufacturers.

D. *Reed 1979*

The "Survey of Biomass Gasifications," authored/edited by Dr. Reed is presented in three volumes. This reference offers a "Synopsis and Executive Summary" in Volume I. There it provides the highlights and summarized the findings of the other volumes. Volume II is the "Principles of Gasification."

This treatment of the principles of gasification is thorough. The survey explains the thermodynmaics and kinetics of gasification reactions as well as pyrolysis. The authors presented in this volume describe the specifics of using biomass as the fuel for gasification. The many different talented and qualified researchers present the technical background necessary for understanding the science, the engineering, and the commercialization of biomass gasification.

Volume III is "Current Technology and Research" as of 1979. However, the status of gasification processes as described is still current enough to be useful. The chapters on economics, gas conditioning, and fuel synthesis address important issues in gasification. This volume also discusses the federal government's role in the development of this technology and makes recommendations for future research.

E. *Tuttle 1978*

In "Combustion Mechanisms in Wood Fixed Boilers, the author published his contribution to the understanding of wood fuel beds. The gasification process described is the fixed bed counter flow system. Gas produced by a pilot plant was sampled and reported. Two aspects of the operation make the data different from prior gas analyses on the fixed bed process. The gasifier was operating at steady-state and at heat-release rates in the range of one-half million Btu per hour per square foot of grate area.

The research was conducted to increase the understanding of combustion occurring in wood fired boilers. The experimental portion demonstrated that particulate emissions decreased and gas quality improved when the depth of the fuel bed was increased.

The theoretical treatment of heat transfer in the fuel bed contributes to the understanding necessary to avoid slagging and overheating of grates or other air distribution systems. The temperature profiles reported are for biomass fixed fuel beds in counter flow. Some of the empirical data were transferred from coal fuel bed research.

The research defines the important differences and similarities between biomass fuels and coals. The author introduces the concept of a fully developed fuel bed. The report contains sections on combustion in the fuel bed, the fuel bed as a gas producer, charcoal and pyrolysis, and on calculating air requirements.

III. GASIFICATION PROCESSES

The gasification processes commercially available for producing low Btu gas from biomass can be separated into three categories. The fluid bed category includes only processes with sand in the bed. The down-draft processes vary. The process commonly called the down-draft is an overfed cocurrent flow. The fixed bed counter flow process is normally an up draft overfeed. It occupies the third category, the fixed bed process.

A. *Fluid Bed Gasifier*

Gases produced by a fluid bed gasifier vary widely depending on several parameters. Fuel moisture, bed temperature, bed depth, gasification rate, sand size, char reinjection, and air temperature and location of fuel inlet all affect the gas composition. Many of these parameters are different by design in gasifiers from different manufacturers. Even though the gasifiers appear nearly the same, the processes occurring in the bed can be quite different. However, there are some common features.

A fluid bed gasifier consists of an air distribution plate or system, a bed of sand several feet deep, a system to feed fuel into the bed, and a vertical chamber of any desired shape. The air is forced through the distribution plate and through the sand at a velocity high enough to fluidize the bed of sand but not high enough velocity to entrain the sand.

The sand must be maintained at a temperature above the ignition point of the fuel. The sand absorbs energy from exothermic reactions of partial oxidation and transfers that energy to endothermic processes of gasification. The bed is a well stirred reactor and the sand is nearly uniform in temperature.

A substantial amount of bed material and unburned solid fuel is entrained in the gases which exit the top of the reactor. Inertial separators are used to remove particles from the gases. The sand must be returned to the bed and the unburned fuel should be recycled.

Methods of fuel feeding are generally propriatory because of the difficulty involved in fuel distribution. Biomass and charcoal from it are too low in density to be part of the well stirred reactor. Instead, the low density fuel particles rapidly elutriate and are entrained. Oxygen in the air oxidizes some of the solid carbon and most of the hydrocarbon vapors. The gas quality is generally lower than other processes produce, and the gas is very hot and extremely dirty. However, most of the heavy hydrocarbons are partially oxidized to gases. The solid particles can be satisfactory removed. The unburned carbon particles can be returned to the bed and oxidized or gasified. The gas temperature can be reduced to a manageable level. The gas produced is clean enough and combustible and in most cases that is what matters most.

Fluid bed gasifiers can be designed to have bed temperatures lower than 1000°F or higher than 1600°F. Pyrolysis of the biomass is rapid and occurs at whatever temperature the bed is maintained. Selecting the optimum bed temperature depends on whether maximum charcoal or maximum methane or minimum charcoal and tar are desired.

In comparing gasification processes, there are three distinguishing features: the air required per pound of fuel, the degree to which the carbon is gasified, and whether oxygen in the gasification air directly contacts the heavy hydrocarbons produced during pyrolysis. The fluid bed process differs from the fixed bed process in these three respects.

The fluid bed process is unique among biomass gasifiers in one important capability. Biomass fuel in any particle size range, any moisture content, and any ash or grit content can be gasified.

B. Down Draft Gasifier

The cocurrent flow, fixed bed gasification process has traditionally been called the down draft gasifier. Although the fuel and the gases do not always flow in the same direction, variations on the process are generally called down draft gasifiers. Based on the three distinguishing features mentioned above, this process resembles the fluid bed. One process injects the air near the middle of the reactor and removes pyrolysis gases from the top while the primary gasification products exist at the bottom. This process resembles the STOIC two stage fixed bed process more than the down draft.

The down draft process permits the pyrolysis products and the gasification air to mix and react. Both pyrolysis products and oxygen are free to react with the charcoal. This process and the resulting producer gas resembles the fluid bed process rather than the fixed bed. In both processes, the temperature is too low for most reactions to occur rapidly, oxidation reactions are likely to predominate. The incandescent char may be as high as 1800°F. Some of the heavy hydrocarbons crack into smaller molecules. The heavy hydrocarbons have a hydrogen to carbon ratio similar to acetylene. When they are cracked, acetylene is produced.

The moisture and pyrolysis products which are mixed with the gasification air can quench the oxidation zone. This process is very sensitive to moisture in the fuel. Fuel moisture below 25% is required.

One variation which protects the oxidation zone from quenching is the injection of the gasification air into the incandescent char from the side of the reactor rather than the top. This does not change the fact that moisture and pyrolysis products must pass through the hot char. The process gains stability, however, because a high temperature reaction zone is maintained near the air inlet. The process is also significantly different.

A step further from the original concept of the down draft process is when the flow of pyrolysis gases are counterflow to the fuel. The gases have been drawn from the top of the reactor and sent to a separate combustion chamber. Better control is gained over the partial oxidation of the heavy volatiles. The hot gases are then returned to the gasifier to continue the path through the incandescent char. The gases may react with charcoal, more likely they serve to moderate the bed temperature.

The air-carbon reaction may occur at a higher temperature than before. Greater conversion of carbon to carbon monoxide may result. Stability is added to the process; gas quality is probably higher. The process in the gasifier begins to resemble the fixed bed up draft process, especially if steam is used to control the bed temperature.

The down draft gasifiers have the strictest fuel limitations. The fuel must be dry as mentioned. The process performs best using a fuel particle size range which will provide passages for the gas. Pieces of fuel larger than an inch in diameter work best. Two-inch cubes are ideal. Fines can not be tolerated to any appreciable extent. Since the oxidation zone is below the melting temperature of ash and grit normally found in biomass, the process will tolerate dirty fuel.

The size of a down draft gasifier is usually limited by the distance air will penetrate when injected from the side of the reactor. Diameter is also limited by the need to support the fuel bed and have a chamber beneath the fuel bed.

Down draft gasifiers have always been attractive for two reasons. The air inlet seals the fuel inlet. Producer gas does not leak through the fuel inlet. Even more important is the concept of passing the products of pyrolysis through the hot charcoal to crack the heavy hydrocarbons.

C. Fixed Bed Updraft Gasifiers

In the fixed bed updraft gasifier, a fuel bed is established as shown in Figure 1. The solid particles are not free to move and mix. If the fuel particles are held in position while passing air through the bed of fuel, the layers seen in Figure 1 develop naturally. As fuel is consumed by drying, pyrolysis, and gasification, fresh fuel is added to the bed. The fresh fuel is added to the top of the fuel bed while the air enters the bottom. The fixed bed process develops and continues.

The fuel bed, though still fixed, develops the layers shown in Figure 2. If the air flow were reversed, the process would convert to the down draft process. The charcoal layer in Figure 2 may reach 1800°F, whereas the oxidation zone in the fixed bed process has been measured at 3000°F.

If the charcoal reaches 3000°F in the fixed bed, the gases may reach 4000°F. At those temperatures, oxygen is consumed by the carbon and all carbon dioxide is reduced to carbon monoxide in a very short residence time. Some of the best producer gas reported (Tuttle 1978) was produced in fuel beds

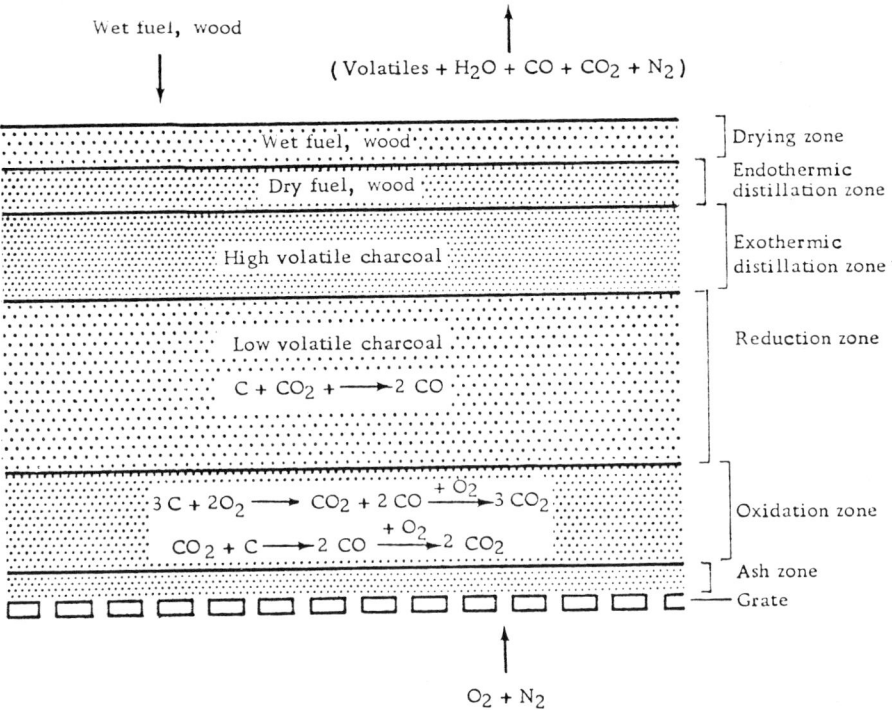

FIGURE 1. Fixed bed, updraft gasification process in a wood fuel bed, a schematic representation.

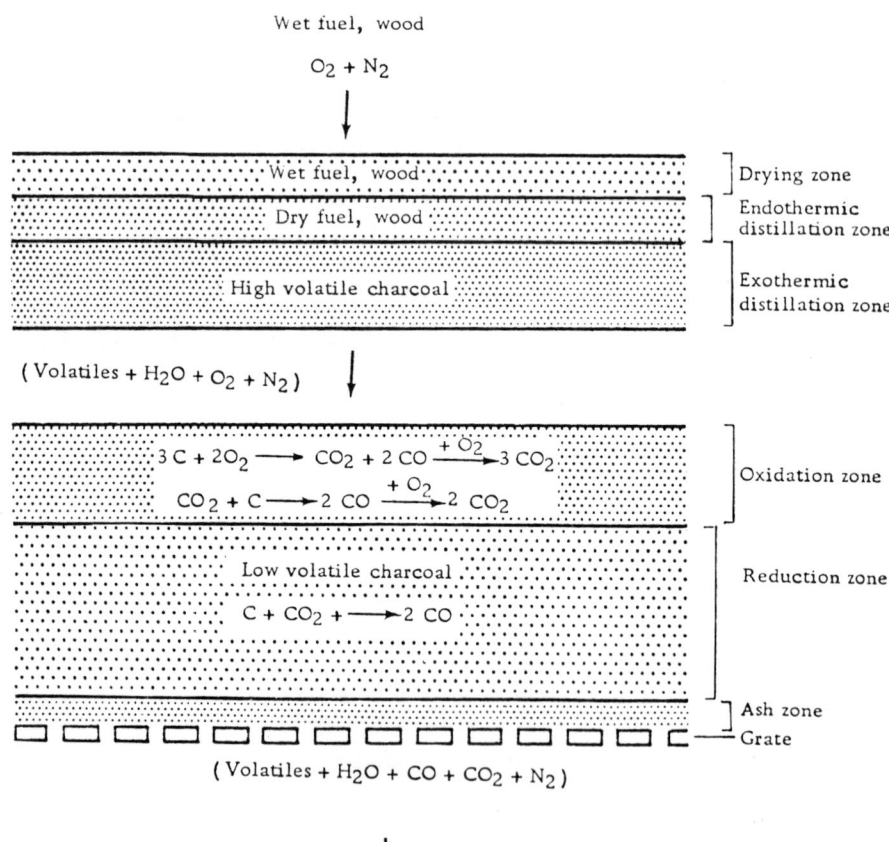

FIGURE 2. Downdraft gasification process in a wood fuel bed, a schematic representation. (From Tuttle [1978].)

four inches thick. The entire fuel bed presented in Figure 1 was only four inches thick. A two-inch fuel bed made only slightly poorer quality gas.

Energy to drive the process comes from oxidation of carbon in the fixed bed. The adiabatic flame temperature for carbon and air going to carbon monoxide is about 2800°F. The reaction rate for reduction of carbon dioxide is very high at 2800°F and above. Essentially all carbon dioxide produced in the oxidation zone is reduced to carbon monoxide. The carbon dioxide reported in producer gas from fully developed fuel beds in fixed bed gasifiers is produced during pyrolysis. The mixture of 2800°F nitrogen and carbon monoxide flows quickly through a layer of dry biomass and destructively distills the volatile matter. Thus, pyrolysis occurs rapidly in the fixed bed process rather than slowly as intuition suggests. Pyrolysis may occur more rapidly in 2800°F gas in the fixed bed than in 1600°F gas and sand in the fluid bed.

The question eventually arises that if temperatures in the vicinity of 2800°T exist in a fuel bed, will the grate or the ash melt? Early research in biomass fuel beds (Tuttle, 1978) developed explanations of the heat transfer mechanisms involved. The main reasons the grate and ash do not melt is because the gasification air convects the energy away as fast as it is transferred to the ash and grate (Tuttle, 1978). The temperature gradient is very steep. One temperature rise, from 150°F to 3000°F, was measured in less than two inches.

The energy contained in those hot gases is used mainly to vaporize the moisture in the fuel. Pyrolysis is believed to be exothermic under these conditions. Gasification in the fixed bed process releases only a fraction of the energy contained in the fuel. A pound of wood requires about six pounds of air for complete combustion and complete release of the chemical energy. In the fixed bed gasifier it takes approximately one pound of air to gasify a pound of wood. To make the fraction of total energy released even smaller, the carbon reacts with oxygen to form carbon monoxide. This reaction releases only about one-third as much energy as complete oxidation of carbon.

Less than a pound of moisture can be vaporized per pound of biomass (dry basis). The exact amount can be calculated for a given fuel. The calculations have never been checked experimentally. If the fuel moisture approaches 50%, moisture removal or an alternate solution should be planned.

Fixed bed gasifiers operate effectively under pressure. However, producer gas tends to leak out through the fuel feed inlet. Rotary valves, tapered screws, and air purge systems can all be made to seal the inlet effectively.

The ash outlet also must be sealed. The ash system can easily be purged with air to prevent gas leakage. However, removing the ash uniformly from across the bottom of the fuel bed is not easy. This was one of the major problems facing gasifier designers. Some operating units are successfully removing the ash.

Channelling, bridging, and blowholes are by far the most pressing problems to be overcome in gasifier design. Theoretically, channelling and blowholes can not occur in certain designs. Bridging can be avoided as well. The theory is supported by the successful operation of at least one fixed bed gasifier which has operated since 1980 without experiencing channelling, bridging, or developing blowholes. This unit is shown in Figures 3, 4, 5, and 6.

The producer gas leaving the fixed bed gasifier is free of ash, dirt and grit and it is relatively cool. The gas has substantially more carbon monoxide than does gas produced by other processes. The heating valve of the gas is about 150 Btu/SDCF with the condensable vapors removed. The gas has more carbon monoxide but less methane, ethene, and acetylene than gas from the other processes. Heating values are comparable. The hydrocarbons normally are burned with the gas. The heating value is effectively two to three times higher.

The condensed hydrocarbons if not burned with the gas create a problem. No market has been developed for the liquids. They cannot be thrown away. They cannot be returned to the gasifier and cracked. The liquids could be sent to a combustion chamber and partially oxidized to form a gaseous fuel and then mixed with the producer gas. However, there are no applications thus far in which that would be an advantage over direct combustion as a liquid fuel.

The gas composition from the fixed bed process is consistent. An increase in the gasification air produces the same gas at a higher rate. The term air/fuel ratio is not important to gas quality. The fluid bed process can very easily change the air/fuel ratio and release more energy per pound of fuel if desired to compensate an increase in fuel moisture. An increase in fuel moisture in the fixed bed does not affect (dry) gas composition because the oxidation, reduction, and pyrolysis zones are insulated from the incoming fuel, Figure 1. The (dry) gas composition changes very little and very slowly to any change in the biomass fuel being supplied.

The same inherent property of the fixed bed can be a disadvantage if the fuel is too wet for too long. An energy balance must be maintained between the energy released during gasification and the energy needed to evaporate fuel moisture If moisture does not leave the gasifier in the gas stream at the same rate it enters with the fuel, then water accumulates in the fuel bed.

The attractive features of the fixed bed process which are different from the other processes have mainly to do

FIGURE 3. *Rome Georgia gasifier. View of the truck dump reclaimer.*

FIGURE 4. *Rome Georgia gasifier. View of fuel storage bin to the left and the gasifier to the right.*

FIGURE 5. Rome Georgia gasifier. Flare gas burner.

FIGURE 6. Rome Georgia gasifer. Tar and oil dual fuel burner with producer gas.

with gas quality. Less nitrogen and carbon dioxide dilute the low Btu gas. The gas is cool and easy to handle. The gas is essentially free of ash, dirt, and grit. The condensable hydrocarbons can be burned, stored as liquids for surge capacity, or sold. However, the most important feature of the gas is its consistent quality. Continuous measurements of carbon monoxide levels have indicated no fluctuations occur.

IV. COMMERCIAL AVAILABILITY OF GASIFICATION

The manufacturers available to design and build gasifiers will be mentioned here only if they have a biomass gasifier in commercial operation. Furthermore, a distinction will be made between gasifiers which have been operating successfully for more than two years and those in startup or otherwise unproven. Gasifiers where existence but not success is known are offered in the latter category. There are some capable manufacturers in that category. In fact, there are two whose pilot gasifiers this author has tested. However, the manufacturers which have had at least one gasifier in commercial operation for at least two years must be treated separately.

A. Units Operating Commercially

Only three manufacturers have commercially operating units. They gasify wood then and pipe the gas to separately controlled burners. One is a fixed bed process by Applied Engineering Company (APCO), Orangeburg, South Carolina. It gasifies whole tree chips at a hospital in Rome, Georgia. The gas is used to fire an old Springfield package boiler producing 20,000 lb/h of steam. Down-rating, particulate emissions, and condensed vapors have not been a problem. The condensed vapors are inertially separated and burned using an oil gun. A second unit is in operation at Florida Power.

The other successful manufacturer is Omnifuel Gasification Systems, Toronto. Their gasifier is a fluid bed process. It fires several burners to heat oil for lumber drying. The gas is cleaned enough using cyclone separators to meet particulate emissions standards. The gas temperature is dropped to a manageable level using a gas-to-air heat exchanger. In this way the gasification air is preheated.

The third manufacturer is Forest Fuels Manufacturing, Inc., Marlborough, New Hampshire. Their fixed bed process is direct coupled to boiler combustion chambers similar to a dutch oven. Although Forest Fuels has had more units operating commercially longer than their competition, less is known about the quality of the gas produced. The feasibility of operating other than direct coupled has not been proven commercially.

B. *Other Experienced Manufacturers*

For a more complete list or additional information regarding manufacturers, the references are recommended, especially Miller, 1983. However, several companies are to be mentioned here.

Two companies which have built commercial fluid bed gasifiers are Energy Products of Idaho and Power Recovery Systems. Energy Products of Idaho has a fluid bed gasifier in startup. Power Recovery Systems, Inc., reorganized from ERCO, has both pilot plant and commercial experience. A third firm, Combustion Power Company, Menlo Park, California, had good pilot gasifier experience and is in the design stage of a commercial fluid bed gasifiers.

Downdraft gasifiers have been applied to biomass fuels with varying degrees of success. Most if not all applications include internal combustion engines. Use of an I.C. engine requires more stringent gas cleanup than does external combustion. The Sermie-Moteur Duvant Company appears to have the most commercial experience. The experience of other foreign manufacturers such as Fritz Werner is acknowledged. However, little first hand knowledge is available.

The University of Missouri-Rolla and the University of California-Davis are conducting the most extensive research on production of low Btu gas from biomass. The pilot gasifier in Rolla is a fluid bed. Commercial technology may eventually develop from that research.

V. SUMMARY

Owing to recent progress and diverse range of existing gasification technologies, this review is needed. Three manufacturers have biomass gasifiers with commercial experience of at least two years. They are Applied Engineering, Omnifuel, and Forest Fuels. Several other manufacturers may be in that category soon. They include Energy Products of Idaho, Power Recovery, Inc., Duvant, Fritz Werner, and Combustion Power Company. Many more manufacturers are in the development stage.

Substantial research has been conducted on biomass fuels and biomass gasification in the past decade. See the recommended references plus research at the Univerity of California and University of Missouri-Rolla.

Three basic gasification processes are being manufactured for low Btu gas from biomass. They are the fluid bed, the down draft, and the fixed bed processes. Each process has advantages and disadvantages compared to each other. The comparative advantages are as follows:

(1) *Fluid Bed Process.* Fuel flexibility in terms of moisture, size and ash content. Tar and char can easily be reinjected. Stable operation.

(2) *Downdraft Process.* Products of pyrolysis pass through hot charcoal which results in most of the tars/oils being cracked or oxidized to gases.

(3) *Fixed Bed Process.* Gas composition in terms of high carbon monoxide, low carbon dioxide and nitrogen, lack of ash and grit, low temperature, valuable liquids, and consistent quality.

REFERENCES

Marenco, Inc. 1982. Wood Gasification/Power Generation Development Porject. Final Report. Alaska Village Electric Cooperative, Inc., Anchorage, March. Conducted for the United States Department of Energy and Alaska Division of Energy and Power Development.

Miller, B. 1983. State-of-the Art Survey of Wood Gasification Technology. Electric Power Research Institute, Palo Alto, May. AP-3101 Research Project 986-9. Prepared by Fred C. Hart Associates, Inc., Washington, D.C.

Oliver, E. D. 1982. Technical Evaluation of Wood Gasification. Electric Power Research Institute, Palo Alto, August. AP-2567 Research Project 986-10. Prepared by Synthetic Fuels Associates, Inc. Palo Alto.

Reed, T. B. et al. 1979. Survey of Biomass Gasification. Vol. I: Synopsis and Executive Summary; Vol. II: Principles of Gasification; Vol. III: Current Technology and Research. NTIS, U. S. Dept. of Commerce, July. TR-33-239, Solar Energy Research Institute (DOE), Golden, Colorado.

Tuttle, Kenneth L. 1978. Combustion Mechanisms in Wood Fired Boilers. Oregon State University, Corvallis.

INDEX

A

Agricultural residues, 64, 70, 80, 90, 92, 95, 165
  almond shells, 80, 90
  bagasse, 5, 14, 31, 32, 42
  cherry pits, 76, 82, 84, 90
  coconut fiber, 76, 82, 84
  coconut shells, 76, 82, 84, 90
  corncobs, 3, 4, 5, 7, 16, 30, 31, 32, 39, 44, 165
  cotton gin trash, 70, 76, 78, 80, 82, 84, 85, 90, 95, 96, 97
  cottonseed hulls, 3, 5, 18, 31, 32, 46
  grape pomace, 71, 76, 78, 80, 82, 84, 90, 95
  grasses, 71
  manure, 165
  mote trash, 76, 82, 90
  olive pits, 70, 76, 82, 84, 90
  orchard prunings, 78, 80, 82, 85, 90
  peach pits, 70, 76, 78, 80, 82, 84, 90, 97
  peanut shells, 3, 5, 15, 31, 32, 43, 76, 82, 84, 90
  pecan shells, 3, 5, 13, 31, 32, 41
  pistachio nuts, 76, 82, 84, 90
  plum pits, 76, 82, 84, 90
  rice hulls, 3, 5, 17, 31, 45, 70, 76, 80, 82, 84, 90, 95, 96, 97
  stone fruit prunings, 71, 90
  sugar beet residues, 95
  tomato pomace, 70, 76, 78, 82, 84, 90, 95
  vineyard prunings, 71, 76, 78, 80, 82, 85, 86, 90
  walnut shells, 76, 82, 84, 90, 97, 170
Alcohol, 165
Alkyl aromatics, 30
Anhydroglucose, 7
Arabinose, 111
Ash, 5, 85, 111, 113, 212, 273
Ash fusion temperature, 75, 79, 86

B

Baghouse, 205
Biomass combustion systems
  fluidized bed, 97
  pile burner, 97
  recovery boiler, 146, 147, 152, 153
  spreader stoker, 97
  suspension burner, 97
  tunnel furnace, 97
Biomass components
  cellulose, 7, 93, 103, 148
  extractives, 111, 148
  hemicelluloses, 7, 102, 138, 148
  lignin, 7, 93, 94, 111, 144, 148, 156, 159
Bisulfate, 129
Bonneville Power Administration, 225, 226
Bulk density, 74, 79, 84, 85

C

Calcium, 113, 206
Calcium sulfate, 111

Carbohydrates, 94, 102, 120, 148, 156, 159
Char, 2, 60
Chemical pulping, 143
Clean Air Act, 245
Cogeneration, 97, 143, 149–152, 186, 189, 191, 192, 195–196, 234
Combustion, 3, 95, 97
Competitive markets, 164, 167, 178, 229
Corrosion, 9–21, 26, 205, 213
Corrosion rates, 9, 27
Corrosivity, 9–21
Cost(s)
  avoided cost, 225
  fixed costs, 175
  marginal cost, 175, 176, 186–189, 201
  opportunity costs, 201
  variable costs, 175
Cotton cellulose, 106

D

Demand
  demand curve, 170, 189
  demand functions, 168, 169, 219
  derived demand, 219
  household demand, 167–170
  market demand, 174–175
  producer demand, 170–174
  quantities demanded, 168
Diesel engine, 57
Dilute acid hydrolysis, 102, 115–118
Distillation, 51

E

Elasticity
  expenditure elasticity, 169
  income elasticity, 169
  price elasticity, 169, 170
Electricity, 70, 186, 194, 197, 219, 224, 232
Electrostatic precipitator, 215
Energy crops, 165, 166
Ethanol, 144

F

Forest management, 221, 233–234
Forestry residues, 64, 70, 165, 166, 221
Fossil fuels
  coal, 87, 88, 89, 90, 92, 154, 164, 176, 215, 221
  diesel oil, 50, 52, 55, 60
  gasoline, 53, 60
  natural gas, 96, 154, 164, 226, 265

oil, 154, 159, 164, 176, 193, 194, 215, 220, 226
Fuel analyses
  higher heating values, 5, 73–74, 82–83, 90–92, 93, 210, 212
  moisture content, 5, 78, 194, 195–197, 269
  proximate analysis, 72, 75–77, 88, 210–213
  screen fractionation, 75, 79, 85
  ultimate analysis, 5, 32, 73, 78, 80, 89
Furfural, 7, 102, 103, 107
  yields, 107–108, 134–138

G

Galactose, 111
Gasification, 2, 4, 9, 39, 264
  downdraft gasifier, 4, 264, 265, 269–270, 272, 278, 279
  fixed bed updraft gasifier, 4, 264, 266, 270–277, 279
  fluid bed gasifier, 264, 266, 267–269, 277, 279
Gasoline engine, 58
Glucose, 111, 120

H

Heat balance, 147, 214
Hydrochloric acid, 118
Hydrogen donor solvents, 29
  alkyl cyclohexanes, 29
  decalin, 28, 29
  methyl cyclohexane, 29, 31, 38, 40, 41, 42, 43, 44, 45, 46, 47
  tetralin, 28, 29
Hydrolyzing solution, 108, 112

I

Imperfect markets, 164, 167, 181–183
Internal combustion engine fuels, 21

K

Kraft pulping, 144, 145, 147

L

Levoglucosan, 3
Lignocellulose, 102, 106
Lignosulfonates, 144
Linear programming, 198–200
Logging residue, 220, 222, 226, 236, *see also* Forestry residues

Index

M

Mannose, 111
Marginal revenue, 181, 183
Molasses, 165
Municipal solid waste, 165, 166

N

Net present value, 220, 231–232, 235, 243
Net public benefits, 231–232, 245, 246, 249

P

Paraffinic hydrocarbons, 30
Perfect competition, 180–181
Petrochemical industry, 63
Phenolics, 4, 9, 29, 62
  cresol, 4, 7
  guaiacol, 4, 7, 29
  phenol, 4, 7, 144
Phenolic acids, 93, 94
Pile unmerchantible material (PUM), 227, 228
Pricing
  price, 164
  price determination, 178–179
  shadow pricing, 200, 201
  transfer pricing, 166, 179–185, 198–200
Process flowsheet, 60–62
Process heat, 97
Process steam, 97, 192, 219
Public Utility Regulatory Policies Act (PURPA), 225
Pulp mills, 70, 152–153, 180, 236
Pulp yields, 152, 153
Pyrolysis, 2, 3, 4, 9, 60, 269, 273

R

Regeneration/reforestation, 221, 242–243
Rome, Georgia, 275–276, 277

S

Salt water, 204
Sawmills, 70, 167, 180, 185, 234, 236
Scrubbers, 204–205, 215
Smoke management, 237–241
Sodium, 206, 207, 208
Soil disturbance, 241–242
Sugar products, 165, 166
Sugar yield, 115
Sulfite pulping, 144
Sulfuric acid, 111, 112, 118, 130
Supply, 175–178
Supply function, 176

T

Tall oil, 144
Tech-Air, 4, 5, 7, 9, 32, 33–37
Timber sales, 228
Torula yeast, 144
Transportation cost, 165, 177
Turpentine, 144

U

U.S. Forest Service, 219, 227, 228, 229, 230
Uronic acid, 109

V

Vanillin, 144
Visibility, 238–239
Volatile organic acids, 4, 6
  acetic acid, 6, 7, 21, 136–137
  butyric acid, 6
  formic acid, 6, 21
  isovaleric acid, 6
  propionic acid, 6

W

Washington water power, 224, 236
Water hydrolysis, 115–118
Wood fuels
  bark, 78, 83, 91, 92, 94, 167, 170, 209
  black liquor, 148, 153, 154, 159
  charcoal, 268
  cull material, 77, 78, 80, 83, 91, 92
  hog fuel, 78, 154, 159, 167, 170
  planer shavings, 78, 170
  sanderdust, 78
  sawdust, 78
  spent pulping liquor, 143
  wood chips, 5, 7, 8, 12, 31, 32, 40, 78, 102, 110, 180
Wood species
  balsam fir, 94
  big leaf maple, 77, 80, 83, 84
  black oak, 77, 80, 83, 84
  black spruce, 94
  canyon live oak, 77, 80, 83, 84
  chinkapin, 77, 80, 83, 84
  common beech, 94
  Douglas fir, 77, 80, 83, 84
  eastern hemlock, 94
  jack pine, 94
  lodgepole pine, 94
  madrone, 77, 80, 83, 84, 86
  Norway spruce, 94

Pacific silver fir, 206, 210, 212, 213
paper birch, 94
ponderosa pine, 83
red alder, 77, 80, 83, 84
red gum, 94
Sitka spruce, 204, 206, 207, 212, 213
slash pine, 94
southern red oak, 102, 109, 110, 111, 138
sugar pine, 94
tan oak, 77, 80, 83, 84, 86
true firs, 207
western hemlock, 77, 83, 204, 206, 207, 208, 209, 210, 212, 213
western red cedar, 77, 83

X

Xylan, 102, 105
Xylose, 102, 103, 108, 120, 139
Xylose yields, 114, 129–133, 139

Y

Yarding unmerchantible material (YUM), 219–220, 227, 228, 229, 230, 231, 235, 243, 244, 245, 246–247, 248